❧ PRAISE FOR ❧
THE ART OF NATURAL CHEESEMAKING

"Finally, the cheesemaking book I've always wanted to read! For anyone interested in traditional, low-tech cheesemaking, *The Art of Natural Cheesemaking* will be an empowering tool."

—SANDOR ELLIX KATZ, author of
The Art of Fermentation, from the Foreword

"If you want to know every possible detail about cheesemaking the natural way and on a small scale in your home, *The Art of Natural Cheesemaking* is the book for you—even if you'd just like to dabble in your kitchen. There are chapters on kefir, yogurt cheeses, and paneer for beginners and, for advanced students, detailed instructions on how to make rennet from the fourth stomach of a calf. Everything is beautifully illustrated and carefully explained. This book will entice many to join the ranks of those engaged in the art of transforming milk to delicious end products. As the old saying goes, 'Blessed are the cheesemakers.' Many more will become blessed thanks to David Asher's work."

—SALLY FALLON MORELL, president, the Weston A. Price
Foundation, and cheesemaker, P. A. Bowen Farmstead

"*The Art of Natural Cheesemaking* is a breakthrough book. The interest among eaters to explore this next stage in do-it-yourself living in the 21st century has finally reached dairy. What's great about Asher's book is that it is practical and zeroes in on cheese products one may actually make successfully at home. It is unlikely that DIY cheesemaking will put any cheesemonger or cheese producer out of business. Quite the opposite. In fact, the more we remove the mystery to manufacturing even the simplest of cheeses at home, the more we will come to admire the craftsmanship that dairy farmers and artisanal cheesemakers bring to their work, to make life better and tastier for the rest of us."

—RICHARD MCCARTHY, executive director, Slow Food USA

"David Asher's book is brave and important, teaching us to tend to what matters by helping us understand process before recipes. This book expands the boundaries of sustainability, deepening the power of independent autonomy and local flavor, making our world more delicious."

—SHANNON HAYES, author of *Radical Homemakers:
Reclaiming Domesticity from a Consumer Culture*

the ART of NATURAL CHEESEMAKING

the ART of NATURAL CHEESEMAKING

Using Traditional,
Non-Industrial Methods
and Raw Ingredients
to Make the
World's Best Cheeses

DAVID ASHER

Foreword by Sandor Ellix Katz

PHOTOGRAPHS BY KELLY BROWN

Chelsea Green Publishing
White River Junction, Vermont

Project Manager: Patricia Stone
Project Editor: Benjamin Watson
Copy Editor: Laura Jorstad
Proofreader: Eileen Clawson
Indexer: Shana Milkie
Designer: Melissa Jacobson

Printed in the United States of America.
First printing June, 2015.
10 9 8 7 6 5 4 3 2 1   15 16 17 18

Our Commitment to Green Publishing
Chelsea Green sees publishing as a tool for cultural change and ecological stewardship. We strive to align our book manufacturing practices with our editorial mission and to reduce the impact of our business enterprise in the environment. We print our books and catalogs on chlorine-free recycled paper, using vegetable-based inks whenever possible. This book may cost slightly more because it was printed on paper that contains recycled fiber, and we hope you'll agree that it's worth it. Chelsea Green is a member of the Green Press Initiative (www.greenpressinitiative.org), a nonprofit coalition of publishers, manufacturers, and authors working to protect the world's endangered forests and conserve natural resources. *The Art of Natural Cheesemaking* was printed on paper supplied by QuadGraphics that contains at least 10% postconsumer recycled fiber.

Library of Congress Cataloging-in-Publication Data
Asher, David, 1980–
  The art of natural cheesemaking : using traditional, non-industrial methods and raw ingredients to make the world's best cheeses / David Asher.
      pages cm
  Includes bibliographical references and index.
  ISBN 978-1-60358-578-1 (pbk.) — ISBN 978-1-60358-579-8 (ebook)
1. Cheesemaking. I. Title.

SF271.A843 2015
637'.3—dc23
                          2015006122

Chelsea Green Publishing
85 North Main Street, Suite 120
White River Junction, VT 05001
(802) 295-6300
www.chelseagreen.com

All good cheese is wild and free.

CONTENTS

Finally, the cheesemaking book I've always wanted to read!

Cheese is such an important realm of fermentation and food preservation. It transforms one of the most perishable of food products, milk, into a food that can be stored and transported, even without refrigeration. And its brilliant diversity of flavors, smells, textures, and appearances exemplifies the creativity and adaptability of human culture. One crucial factor in this diversity is microbial biodiversity, in that much of the variation in cheeses is due to different types of organisms and microbial communities, as selected by aging environment and surface treatment.

Most of the contemporary practice of cheesemaking, along with most of the contemporary literature, relies on the use of laboratory-derived pure-strain cultures, which are a relatively new phenomenon, possible for barely 150 years and widely used for a much shorter time. These pure-strain cultures represent a radical departure from the entire earlier history of cheesemaking, in which all cheeses relied upon the rich microbial communities indigenous to milk, often accentuated by the practice of backslopping, which means incorporating a little of the previous batch into each subsequent one.

While cheese is such an extraordinary manifestation of biodiversity, the logic behind most contemporary cheesemaking is that of monoculture, using a single microbial strain (or possibly two or three) in lieu of the broader microbial communities used traditionally. This is as true for most farmstead, home, or hobby cheesemakers as it is for the mass producers.

The Art of Natural Cheesemaking reclaims this earlier cheesemaking legacy, with great practical information on how to work with raw milk, along with widely available kefir grains or yogurt, to encourage different types of microbial communities for different types of cheeses. David Asher poses "a challenge to conventional cheesemaking," as he puts it, and offers a bold vision of cheesemaking as wild fermentation, by revisiting older methods that rely upon the biodiversity of raw milk. "We need to stop thinking of the cultural regimes in cheese as individual strains of bacteria or fungi," he declares. "It is the reductionist industrial approach to cheesemaking that has singled out what are believed to be the important players in an evolving cheese; the minor players have been cast aside and forgotten. Yet it's the combination of the important players and their supporting microorganisms that make for a healthy cheesemaking. For it is not an individual species that makes a cheese, but a community of microorganisms . . ."

The Art of Natural Cheesemaking starts with this impassioned and compelling manifesto; then the rest of the book is devoted to methods, in sufficient detail to guide the reader through the process. I have spent time with David, as both teacher and student. I have greatly enjoyed his delicious cheeses, and I've had success using those of his methods that I have tried. He provides clear and methodical instructions, along with great illustrative photographs, to guide you through making an incredible array of cheeses in your home kitchen. And he does so relying entirely upon microbial communities indigenous to the milk, or those of yogurt or kefir. He offers simple techniques, emphasizing improvisation with what's easily available rather than buying specialized gadgetry.

This book appears in the context of a broader fermentation revival, in which diverse ancient fermentation traditions are being reawakened and explored anew after a period of eclipse by industrial production. Home practitioners and small local enterprises alike are experimenting with traditional techniques. For anyone interested in traditional, low-tech cheesemaking, *The Art of Natural Cheesemaking* will be an empowering tool. Welcome to the fermentation revival!

Sandor Ellix Katz
March, 2015

PREFACE

I'm visiting this remote island off the west coast of British Columbia on a writer's retreat organized by a stranger. I received an e-mail from Jaylene out of the blue, inviting me to come to Lasqueti to teach a cheesemaking class. She asked me what she could offer to get me to come out and engage her community, far off the beaten track. I told Jaylene that I had wanted to visit her island since I'd first learned of it, and that I had been waiting for years for an invitation to come. All that I'd ask, I added, would be for a quiet cabin to write in . . .

Lasqueti is different. It is a place where feral sheep have the run of the island; where severe storms born of the Pacific crash ashore with ferocious winds; and where Douglas firs, normally majestic broad-branched trees, cling to the rocky soil and grow twisted and bonsai-like, prostrate against seaside cliffs. It is a maverick island where the community does things their own way, preferring to remain off-grid and off-ferry, preventing the development that has scoured neighboring lands. And it is a place where residents grow gardens, wildcraft, and seek out natural ways of living, and where DIY is a way of life.

It is places like these that truly appreciate my teachings—for it is places like these that have inspired my teachings. It is on the islands of the wild west coast of Canada that I learned to farm. And it is on these islands that I taught myself a new way to make cheese.

It is the land, it is the communities, and it is the people of the islands that have inspired my ideals of a natural cheesemaking. Without the healthy ecologies of the farm, the forest, and the sea; without the communities' self-sufficiencies; without the farmers' spirit; and without the people's rejection of the status quo, I would never have put together these thoughts.

The ecologies on these islands are vibrant and alive. You hear them first thing in the morning with the dawn chorus of birds. You see them firsthand walking through mature forest stands. And you taste them, too—the vitality of a healthy soil ecology coming through in the fine flavor of locally grown produce.

These island communities are more introspective, observing the comings and goings more acutely than mainland communities. It is clearer here how tenuous our dependency on an overextended food system is, as nearly every morsel of food in the grocery store arrives on a truck from "off-island."

The island awareness encourages people to be more invested in what sustains them, and organic gardens are prevalent here as nowhere else in North America. Farmers cultivate diverse farming operations, featuring livestock, row crops, and orchards; compost-building plays a key role in the development of healthy soil; seed-saving practices adapt crops to the local growing conditions; and permaculture systems ease the load on the land. Farmers here also slaughter and butcher their own animals because, though local regulations restrict such activities, it's the only way their isolated farms and communities can remain viable.

The people of these islands support their farmers, who are considered heroes and celebrities, sharing ranks with visiting health-care practitioners and volunteer firefighters. Residents here question authority and the status quo and strive to make the change they wish to see in the world.

It is thanks to the people here, like Jaylene, that I am encouraged to teach. It is thanks to the many organizations here that are promoting food sovereignty and provide venues for my classes that I'm able to teach; and now, it is thanks to Chelsea Green that my words will teach a far greater audience than I could possibly reach.

Lasqueti Island, British Columbia, Canada
April, 2014

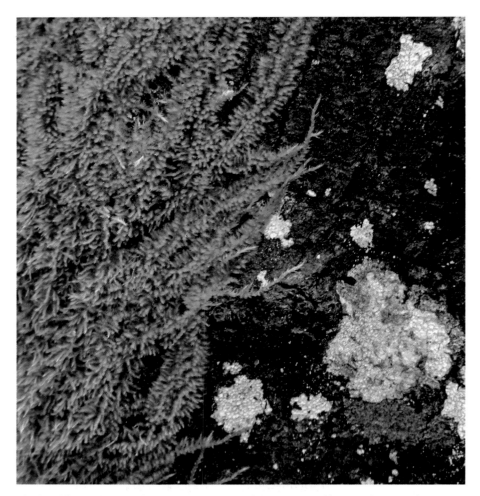

The healthy ecologies of the West Coast are the inspiration for my cheesemaking.

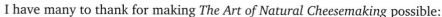

ACKNOWLEDGMENTS

I have many to thank for making *The Art of Natural Cheesemaking* possible:

I couldn't have written this without the authors of my favorite books and websites on fermentation and cheesemaking, who provided valuable information for my cheesemaking experimentation. Thank you, Ricki Carroll, Gianaclis Caldwell, Dominic Anfiteatro, David B. Fankhauser, and, most importantly, Sandor Katz, whom I must also thank for writing the foreword. As well, I'd like to thank the many researchers studying the microbial community of raw milk, cheese, and kefir, for helping me to understand the forces at play in milk's many transformations.

Thank you to all the organic farmers, gardeners, and permaculturists that inspire me, but especially to those who've taught me so much of what I practice: Ron Pither of Varalaya; David Buckner, Adam Schick, Oliver Kellhammer, Liz Richardson and all the stewards at Linnaea Farm; Don and Shanti McDougall of Deacon Vale Farm; Walter Harvey and Lauren Sellars at Snowy Mountain Farm; Nikki Spooner; and to Alyson Chisholm for showing me the Way of Cheese.

Thank you to the farmers of the Hardscrabble Collective for doin' it— Christa Morin and Jeff Hochhalter, Scott Magee, and Emily Adam. And to all of the WWOOFers who have toiled alongside me in the fields.

Thank you to all the milking animals in my life that have provided me with so much goodness to work with: Sunday and Wilderness; Nicky and Toni; and Dorothy, Harold, and Maude.

To the Canadian raw milk farmers who put so much on the line to make good milk available to the public: Michael Schmidt, Alice Jongerden, and many others whom I greatly respect. Your concern for the well-being of the land, your livestock, and the lives of your milk drinkers is unparalleled—thank you.

Thank you to all the artisan cheesemakers out there who inspire me with their beautiful cheeses. Though I am critical of many contemporary cheesemaking practices in this book, I truly respect your work.

Thank you to the food and farming not-for-profits that are changing the way people relate to their food: in particular, Santropol Roulant—for forging my food and community ethic; the student-run UBC Farm for helping me find my inner farmer; the numerous other organizations that have given me the opportunity to teach; and the Slow Food Movement for working to preserve traditional foodways and flavors—your writings are an important motivation for the work I do.

To my many students over the years, especially those who asked the challenging questions and forced me to further my understanding of the cheesemaking process. Thank you for your interest, and for putting my techniques and my teaching to the test.

◇ ◇ ◇

For providing safe spaces for writing, thank you to Mark Lauckner and Leanna Boyer; Jaylene Scheible; Renata Crowe and Dean Sharr; Helen and John O'Brian; and Jagtar Sandhu (as well as the water buffalo farmers of Punjab and Haryana).

For contributions to the photos of the book in the form of farm produce (almost everything in the book is locally grown), props, photo-shoot sites, and space for aging my cheeses, thank you to Molly Wilson and Zack Hemstreet of Bullock Lake Farm; Cliff Leir of Fol Epi; Melissa Searcy; Alastair Heseltine; Kate Schat; Sofya Raginsky and Colin McNair; Mark and Leanna again; Jamie Ferguson; Jan Steinman; and the Gulf Islands of British Columbia, for being so beautiful. However, I cannot acknowledge this place without recognizing that this land is the unceded territory of the Coast Salish First Nations.

◇ ◇ ◇

I cannot offer enough thanks to Kelly Brown for casting my cheeses in their best light. I couldn't have asked for a more considerate photographer.

Thank you to my editor, Ben Watson; my project manager, Patricia Stone, and all the staff at Chelsea Green for putting it all together. It's an honor to have the opportunity to work with such an acclaimed publishing house—thank you for choosing me, an unproven author and unknown cheesemaker!

Thank you to my entire family for their love and support: to Eva, who will never get to read this book she so encouraged; to my grandmother Balka for keeping traditions alive; to my brother and sister, Jeremy and Marla—for being the best brother and sister possible; and to my mother and father, Mira and Jack, for all of your support—but especially for digging up a small plot in your yard and showing me how to plant seeds and care for a garden . . . and for not giving me too much hell when I dropped out of engineering school to become a farmer.

And finally, thank you, Kathryn. For being so wonderful, and for making room in your life for my stinky cheeses . . . I love you!

the ART of NATURAL CHEESEMAKING

Cheesemaking, as practiced in North America, is decidedly unnatural.

Is there an approach to the art that's not dependent on packaged mesophilic starter cultures, freeze-dried fungal spores, microbial rennet, and calcium chloride? Do cheesemakers really need pH meters, plastic cheese forms, and sanitizing solutions? Are modern technologies the only path to good cheese?

What of traditional methodologies? Did cheesemakers make consistently good cheese prior to pasteurization? Did cheeses fail if they weren't made in stainless-steel vats with pure strains of *Lactobacilli* and triple-washed surfaces? Where are the guidebooks that teach traditional methods? Have our ancestors' cheesemaking practices been lost to the forces of progress and commercialization?

I believe that the quality and taste of cheese have declined dramatically as traditional methods have been abandoned. And that the idea—propagated by the industrial cheesemaking paradigm—that traditional ways of making cheese, with raw milk and mother cultures, make for inconsistent and poor-quality cheese is a myth. For there is wisdom in the traditional practices of cheesemakers . . .

Generations upon generations of traditional cheesemakers evolved the diverse methods of making cheese while carefully practicing their art. All classes of cheese were discovered by cheesemakers long before they had a scientific understanding of the microbiological and chemical forces at play in its creation. Industry and science hijacked cheesemaking from the artisans and farmers some 150 years ago, and since then few new styles of cheese have been created; yet during that time hundreds, possibly thousands, of unique cheeses have been lost.

Standard methods of cheesemaking—reliant on pasteurization, freeze-dried starters, and synthetic rennets that interfere with the ecology of cheese—are equivalent to standard practices in industrial agriculture, such as the use of hybrid seeds, chemical fertilizers, and pesticides that have overtaken traditional agriculture, and conflict with the ecology of the land. Cheese comes from the land and is one of our most celebrated foods; yet its current production methods are environmentally destructive, corporately controlled, and chemically dependent. In its eating we're not celebrating the traditions of agriculture but rather pasteurization, stainless-steel production, biotechnology, and corporate culture. If we gave its methods of production some thought, we wouldn't want to eat the stuff!

It strikes me as absurd that there is no commonly practiced natural cheesemaking in North America. Farmers practice ecologically inspired agriculture; brewers are making beers and wines with only wild yeasts; bakers are raising breads with heirloom sourdough starters; and sauerkraut makers are fermenting their krauts with only the indigenous cultures of the cabbage. But cheesemakers are stuck in a haze of food technology, pasteurization, and freeze-dried commercial cultures, and no one even questions the standard approach.

Other cheesemaking guidebooks insist that home cheesemakers adopt the industrial approach to cheese along with its tools and additives. Their advice is based on standards put in place to make industrial production more efficient, and a mass-produced product safer. But for small-scale or home-scale cheesemaking, a different approach can work.

A Different Approach

From the making of my very first Camembert, I knew there had to be a better way than the cheesemaking methods preached by the go-to guidebooks. I just couldn't bring myself to buy a package of freeze-dried fungus, and my search for alternatives to commonly used cheese additives led to a series of discoveries—about the origins of culture, about the beauty of raw milk, and about the nature of cheese—that set in place the philosophies of this guidebook.

Not being one to blindly follow the standard path, I set out to teach myself a traditional approach to cheesemaking. The methods I share in this book are the result of 10 years of my own experimentations and creative inquiry with milk: years of trial and error in my kitchen, rediscovering, one by one, a natural approach to making every style of cheese.

I now practice a cheesemaking inspired by the principles of ecology, biodynamics, and organic farming; it is a cheesemaking that's influenced by traditional methods of fermentation through which I preserve all my other foods; and a cheesemaking that's not in conflict with the simple and noncommercial manner in which I live my life. I now work with nature, rather than against nature, to make cheese.

When I teach my methods to students, there is not a single book that I can recommend that explores a natural cheese philosophy, and no website to browse but my own. It is this absence of information in print and online that led me to write this book. I never thought that I'd be an author, but I felt compelled to provide a compilation of methods for making cheese differently. For it's about time for a book to lay the framework for a hands-on, natural, and traditional approach to cheese.

The techniques presented in this book work. And the photographs within, featuring cheeses made by these methods, are the only proof I can offer. I wish I could share my cheeses with you so that you could taste how delicious a more naturally made cheese can be, but unfortunately I cannot sell the cheeses I make because raw milk and food safety regulations restrict me from selling cheeses made in the small-scale and traditional manner that I practice. If small-scale and traditional practices are constrained by regulations controlling cheese production and access to raw milk, perhaps it is time to question the authority of these standards.

We need a more radical cheesemaking, a more natural approach to the medium of milk. But it's surprising that it's come to me to lay this foundation; for who am I, but a small farmer and a humble cheesemaker . . .

More About Me

I am an organic farmer, first and foremost. The agriculture I practice is my tool for positive change; my right livelihood. Out of a strong environmental ethic I chose this vocation. And my work feeds both my soul and my community.

For 10 years now I've been working on small organic farms, bringing produce to local farmers markets, offering a CSA program, and raising goats, in part to help provide fertility to the soil I cultivate, but also for their precious and nourishing milk.

The important responsibilities of compost-building, cover cropping, and keeping livestock to provide fertility form the basis of my organic philosophy. Crop rotation and rotational grazing keep the fields and livestock happy. And *Hügelkultur* (raised beds layered with soil, organic matter, and woody debris), chicken tractors, and interplantings all make the farm a very productive operation.

I have dedicated a large section of my garden to saving seeds from open-pollinated varieties of vegetables, varieties that are in constant adaptation to the growing conditions on the farm. I avoid hybrid crops, as they are costly and corporately controlled and cannot be maintained true to type. And by the mandate of organic agriculture, genetically modified (GM) crops are strictly forbidden.

In my farming practices I strive for a full-circle, organic approach. It's only natural that in my cheesemaking I would strive for a full-circle, organic cheese.

My Fermentationism

My natural approach to farming has influenced the ways I preserve the harvest. I have turned my back on freezing and canning, two methods that are material- and energy-intensive, and have taken up natural fermentation to preserve the fruits and vegetables of my labors.

Every fall, I prepare big batches of sauerkraut made from my cabbages. Crocks bubble over with white winter cabbages, chopped up, sprinkled with sea salt, and flavored with wild-harvested juniper berries. Fluorescent pink krauts, made with red cabbages and purple carrots, provide essential contrast.

My cellar is filled with carboys of mellowing apple ciders and wild blackberry wines. No packaged yeast is added to make these ferments—the natural bloom on the skin of the fruit comes to life in the juice and provides all the yeasts a cider needs to blossom. The beers I brew are fermented with live cultures backslopped from my ciders; the natural fermentation that results gives the beers a beautiful fruitiness, and flavors and acidity reminiscent of Belgian lambics.

The bread upon which I spread my cultured butter is fermented with a sourdough starter that I summoned from its flour. A regular feeding of wholemeal flour is all that the sourdough needs to be healthy and vital. Fermented with diverse cultures, these breads are delicious and nourishing in ways that commercial bread made with packaged strains of *Saccharomyces* yeasts could never be.

And when I make cheese, I strive to make it as naturally as my bread, my cider, and my sauerkraut.

My Cheesemaking

To help preserve my milk I toss in my kefir grains. These little beings ferment the milk, confer upon it a complex community of beneficial microorganisms, and transform it into thick and delicious kefir; once the milk has thickened, the kefir grains are put into a fresh batch of milk to make more kefir. Keeping kefir grains is as simple as feeding them milk every day: A regular feeding helps keep the grains happy and healthy and resistant to invading bacteria and fungi.

Kefir is an important inspiration for my cheesemaking, and the source of my cheesemaking culture. The kefir grains that sustain the culture of kefir are a stable collection of milk-loving bacteria, yeast, and fungi that have been kept for many thousands of years, passed down from generation to generation.

The profile of cultures within kefir grains is remarkably similar to that of raw milk and likely evolved from it. And using kefir as a starter culture in cheesemaking is akin to planting a seed of healthy and vibrant raw milk culture in cheese.

The simple beauty of kefir culture has led me to seek out methods of keeping other cultures useful to cheesemaking. By studying traditional cheesemaking practices, I've discovered methods of keeping the blue cheese fungus *Penicillium roqueforti* on sourdough bread. Traditional smeared cheeses cast light on how *Brevibacterium linens* bacteria can be transferred from one ripe cheese to another. And my kefir, when forgotten once upon the counter for a couple of weeks, grew a beautiful coat of white fungus that led me to realize that kefir grains are an excellent source of the *Geotrichum candidum* fungus, native also to raw milk, that ripens Camembert and other bloomy-rinded cheeses.

At home I make a broad range of cheeses naturally. I regularly prepare fresh cheeses such as chèvre and mozzarella without purchasing any ingredients from cheesemaking supply stores. My cheese cave is filled, seasonally, with aging blue cheeses, washed-rind cheeses, white bloomy-rinded cheeses, cheddars, Goudas, and Tommes, all made naturally with the cultures I keep and good raw milk. And alongside

Kefir grains sustain the culture of a natural cheesemaking.

all of them are the bacterial and fungal "counter-cultures" that give life to my cheeses.

I do not claim to be a professional cheesemaker. I do not make cheese for a living; in fact I cannot even sell my handmade cheese. Instead, I share my contraband cheese with friends and family and advocate for a natural cheesemaking with the Black Sheep School of Cheesemaking.

The Black Sheep School of Cheesemaking

The Black Sheep School of Cheesemaking is my educational endeavor. It is not a brick-and-mortar institution, but a traveling cheese school that offers cheese outreach at community farms, co-ops, and food-sovereignty-minded organizations near and far.

The diverse organizations that I partner with work to create educational gardens, whip together healthy meals-on-wheels, and donate good food boxes for low-income families. They offer classes for farmers and gardeners, create links between consumers and producers, and reconnect children to the foods that sustain them. And they are well respected within their locales for offering innovative examples of food-centric programming that are making their communities stronger.

Teaching cheesemaking is my bread and butter. But the workshops don't just supplement my meager farming income—they also support the organizations that host my classes. Our classes help fulfill the educational mandate of these groups, and their proceeds help these folks continue their good work building a more just and resilient food system.

The Black Sheep School of Cheesemaking in session.

Milk is the origin of nearly all cheesemaking cultures. Here clabber ferments spontaneously from unpasteurized milk.

CHAPTER ONE

A Natural Cheesemaking Manifesto

G ood milk, rennet, and salt. Together with your capable hands, and the cool and humid environment of an aging cave, these are the only ingredients needed to make good cheese. Several other additions can improve results: chestnut leaves or other greenery, ash, wine, and even moldy bread!

Cheesemaking as practiced today in North America has a much longer list of ingredients, including dozens of different strains of packaged mesophilic and thermophilic starter cultures, freeze-dried fungal spores, microbial and genetically modified rennets, calcium chloride, chemical sanitizers, and harsh nitric and phosphoric acids; also, that most important ingredient (which is actually an anti-ingredient), pasteurization. None of these, however, is a necessity for making good cheese!

This book lays the framework for a more natural cheesemaking (see appendix C for a summary), one whose ingredients are simple, whose culture derives naturally from milk, and which is practiced in conditions that are clean but not necessarily sterile, because the cultures are strong and diverse and the cheeses made well.

In this book you will learn to make all cheeses, not just aged ones, with raw milk. You will also gain insight into how to prepare and keep all the ripening cultures a cheese needs, methods that include cultivating white Camembert rinds naturally, growing blue cheese fungus on sourdough bread, and smearing washed-rind cultures from cheese to cheese.

The book demonstrates how to make your own rennet; how to fashion your own cheese forms; how to improvise a cheese press; and how to really use your hands to make cheese. It will instruct you on how to avoid the use of unnecessary additives and questionable ingredients; how to avoid the need to sanitize and sterilize; and how to limit exposure to plastic, a most unnatural ingredient. It presents a down-to-earth and accessible type of cheesemaking, and a traditionally inspired, yet increasingly countercultural approach to the medium.

The Art of Natural Cheesemaking will show you how to take back your cheese.

A Challenge to Conventional Cheesemaking

The methods described herein challenge the beliefs of the conventional cheesemaking paradigm. It is dogma among most cheesemakers that the culture of cheesemaking must come from a package; others believe that cheese cultures evolve from the environment in which cheese is made—including the vat and the cave. But I believe that all of the cultures that make cheese possible are present in its milk . . .

that the cultures of the vat and the cultures of the cave (and even the cultures in the package) all have their origins in the microbiodiversity of raw milk. And my cheesemaking practice confirms this.

A comparison of the modern method of making Camembert with my more traditional method of making this cheese illustrates the fundamental difference of my approach. Contemporary Camemberts are made by inoculating milk with packaged fungal spores that help them develop white rinds. But raw milk can be contaminated by wild fungal spores, which can cause a Camembert to ripen into a blue cheese, so pasteurizing the milk prior to cheesemaking is considered essential. Furthermore, any fungal spores from the air or the tools that contaminate the cheese can also turn it blue, so a sterile cheesemaking process is also justified. A traditional Camembert, however, made with raw milk and washed with whey during its first week of aging, will grow an even coat of white fungus because the conditions created by its handling limit the growth of unwanted blue fungus and encourage milk's native white fungus to flourish, even in an unsterile setting.

Understanding that the culture of cheese evolves from milk changes everything. That these microorganisms are meant to be in milk, and are not just there because of environmental contamination, is food for thought. That these cultures can come to define the development of different ripening regimes simply through the different ways that the milk and cheese are handled is both a very old and a very new idea. After all, cheesemakers have been making their cheeses in the manner that I practice for thousands of years.

The Loss of American Cheesemaking Culture

North America does not have a healthy cheesemaking culture. Our modern methods of cheesemaking are based on fear: fear of raw milk, fear of foreign bacteria, and fear of fungi. Our milk is mistreated and mistrusted; it is stripped of its life through pasteurization, and monocultured strains of laboratory-raised commercial cultures are added to replace its native

cultures in an attempt to create a more controlled, predictable, and presumably safer cheesemaking.

The consequence of this overly controlled approach is that cheese *must* be made in entirely sterile environments; if precautions aren't taken, the sensitive packaged cultures can fail. All of the equipment, tools, and surfaces that come in contact with cheese must therefore be sanitized to avoid contamination. Cheesemakers hardly even touch the cheeses they make; and when they do, their hands are gloved!

The only approach we know is this standard industrial approach: a top-down, corporate- and government-controlled cheesemaking. But it's not just the way industrial cheeses are produced; our "best" home cheesemaking practices are modeled after standard industrial cheesemaking practices. And even the ideal of the American farmstead cheese is to some extent a facade, for though they source excellent quality milk, and strive to make cheese according to traditional methods, artisan cheesemakers operate under the same standards of production and use many of the same ingredients and packaged cultures as large-scale industrial cheesemakers. With their synthetic rennets manufactured in bioreactors in Colombia, and bacterial cultures raised in laboratories in Denmark, how can artisan cheeses truly be considered handmade and local?

We don't make cheese at home in our culture, and it is in part because of the industrial manner in which our cheesemakers practice this art that our cheesemaking culture has been lost. Our standard methods of cheesemaking can never be widely adopted, for they are too challenging for most home cheesemakers to master. Many folks who wish to take up the art find the methods taught to be difficult to follow, expensive, and discouraging. This hands-off approach to cheesemaking is unnatural, un-intuitive, unattractive, and indicative of just how far cheese has strayed from traditional practices.

Even the very cultures of cheesemaking—the bacterial and fungal cultures that make our cheeses—are increasingly under corporate control. The largest

The different ripening regimes (clockwise from top: *Penicillium roqueforti*, *Brevibacterium linens*, and *Geotrichum candidum*) can all be encouraged naturally without the use of packaged cultures.

cheesemaking supply manufacturer in North America and the source of the majority of the cultures used by cheesemakers is owned by none other than the chemical and agribusiness giant DuPont.

The reality that all the cheesemaking cultures used in North American come from packages is symbolic of the state of our cheesemaking culture, and of our culture in general. Like so many other aspects of the way we live our Western lives, we are no longer participants in our culture but are relegated to consumers. We have lost control of the culture of our cheese.

Cheese Sovereignty

Rather than following the standard fear-based cheesemaking approach, I propose a cheesemaking that celebrates life and diversity, and that works with the nature of milk to make good cheese. Only such an approach can form the basis of a truly farmstead cheesemaking, and only such a cheesemaking can establish a healthy cheesemaking culture.

Cheese need not be controlled by corporate cultures. The bewildering selection of packaged cheesemaking cultures doesn't hold a candle to the diversity of beneficial cheesemaking cultures found in raw milk or kefir. Every cheese can be made with the culture of raw milk, yet culture houses insist that dozens of different packaged cultures would be needed to make them all.

Cheeses, even fresh ones, can be safely made with these raw milk cultures. Raw milk cheeses are protected by the many layers of life within them. The diverse bacterial and fungal cultures act as a sort of immune system that restricts the growth of unwanted microorganisms. Cultivating indigenous raw milk microorganisms through traditional and natural cheesemaking practices imbues cheeses with a protective halo that restricts unwanted and possibly pathogenic bacteria—and makes a more relaxed cheesemaking possible.

If eating is a political act, then cheesemaking is even more so. But how can the act of making cheese take a stand against corporate culture when corporations control the very culture that makes our cheese? Making our own cheese naturally assures us that the ingredients used and the processes involved are up to our exacting standards. Making cheese reconnects us with the land, the livestock, and the farmers that feed us. And it can reduce our dependence on an often unjust, inhumane, and ecologically destructive food system.

Practicing a natural cheesemaking not only encourages a more responsible cheesemaking but also promotes a more natural and ecological dairying. Cheese is intimately linked with the milk that makes it, and a more natural cheesemaking depends on a less processed, more ecologically produced milk.

Unfortunately, in most of the jurisdictions where this book is sold, there are severe restrictions on consumer access to the good milk that helps make a natural cheesemaking possible. The rules and regulations that limit raw milk distribution and the commercial production of raw milk cheeses are justified for industrial production, as risks rise dramatically as the scale of an operation grows. But for smaller-scale operations, the risks are considerably less, and a raw milk cheesemaking can be safely practiced. And certainly, for home-scale cheesemaking the slight risks can easily be swallowed!

Such a culture of cheesemaking suffers because of these restrictions on obtaining good milk. How many other artisans, artists, or agriculturalists are forbidden from obtaining the finest materials they could use in their work? We need greater access to good milk to make a natural cheesemaking more accessible.

Cheese is an agricultural product whose making belongs at home and on the farm. It is one of our most nourishing and delicious foods and a celebration of diverse cultures and agricultures from around the world. Through a more hands-on, grassroots, and democratic cheesemaking process, we have the power to preserve what sustains us. It's up to us to preserve the culture of cheesemaking.

Using This Book

This book was originally envisioned as a companion to my in-class teachings.

Hands-on workshops are my preferred method of engaging the public; but without a guidebook that shares the methods I teach, my students were at a loss for learning when they left the class.

Though it's not quite the same as witnessing first-hand the magic of cheesemaking in a class, I've done my best to provide visualizations here of the processes at hand, and descriptions of the steps that best represent the evolution of milk into various cheeses. In some ways, though, this book offers more than my workshops: It ensures that everything I wish to share with my students is as clear as possible, and that no lesson (and no cheese) is left unturned. There's lots to learn in this book—more than I could ever share in a class.

This book is meant as a guide for novice cheesemakers. However, even cheesemakers with some experience may appreciate the different approach to a familiar subject. Established commercial cheesemakers may also find insight into traditional practices that can improve the wholeness of their cheesemaking operation and the flavor of their cheeses. I've tried my best to address all three audiences—but my voice usually leans toward the beginner!

This book outlines the details of a natural cheesemaking practice, which is drastically different from the standard North American approach to the medium. If you've made cheese before according to standard industrial practices, you might need a reeducation to let go of some beliefs about milk and cheese before you can trust the methods of this book.

The methods rely on biodiverse cultures in raw milk or kefir that are adaptable to the different conditions responsible for each cheese's evolution. It is a simplified and more intuitive style of cheesemaking, which you may find easier to follow than the standard methods. With the right understanding of the processes involved, and following the techniques outlined in this book, you can ensure that the cultures you nourish are the ones that come to define the development of your cheese.

All that being said, this book is not just for cheesemakers. Cheese lovers could stand to learn a thing or two about how their favorite food is made. And ethical eaters will find insight in the book's focus on the social and ecological costs of standard cheesemaking practice and alternatives to the status quo—perhaps you might even be inspired to try your hand at making cheese yourselves!

The Flow of Chapters

This book lays out traditional and natural methods for making approximately 30 different cheeses, as well as yogurt, kefir, and cultured butter. My choice of cheeses is meant to be as broad a selection as is needed to explain the different methods of making diverse cheeses without overwhelming the reader; as a result, I've left out many famous cheeses because their making was simply too similar to that of other cheeses. The 30 or so that I describe are divided into 16 chapters, each of which explores a different class of cheese.

To start the book there are six chapters that explore the basic principles of making cheese naturally. To begin, I recommend reading these basics chapters. Chapters 2 through 6 (Milk, Culture, Rennet, Salt, and Tools) cover the background information that will help you understand how milk evolves into cheese. When you feel ready to tackle aged cheeses, read chapter 7, The Cheese Cave, to learn how to handle your cheeses to encourage them to ripen well.

Chapter 2: Milk will help you understand what milk to choose when making cheese. The chapter explores the differences (from a cheesemaking perspective) between raw and pasteurized milk and offers advice on how to source good milk. **Chapter 3: Culture: The Ecology of Cheese** provides a background on the many different microorganisms that live in cheese and an understanding of how to choreograph the natural cultures that define certain cheeses. **Chapter 4: Rennet** will help you understand the use of this coagulating enzyme, and will provide insight into its many different varieties, as well as how to make your own. **Chapter 5: Salt** provides information on how to use this most natural preservative to help cheeses blossom to their full potential. **Chapter 6: Tools** will help you choose the appropriate tools for making cheese. (Hint: You've probably already got all you need at home.) And **chapter 7: The Cheese Cave** offers advice on how to cultivate a cheese aging space, and how to care for cheeses as they mature and ripen.

Once you've read the basics, start with fresh cheeses. Get some experience making unaged cheeses such as yogurt cheese and chèvre before attempting to make an aged cheese. And when you're ready to make an aged cheese, try ones that ripen quickly and don't need a cool and humid cheese cave, such as Dream Cheese in olive oil, feta, or Mason Jar Marcellin, all of which can be aged in a few weeks in your home refrigerator. But don't just jump ahead to the recipes! Be sure to read the full chapter beforehand to better understand the processes involved in the making of each class of cheese.

The easiest cheeses to make are generally toward the beginning of the book, while the more challenging cheeses are found at the end. The simplest cheeses—kefir, yogurt cheese, and chèvre—are explored first, while rennet cheeses and aged cheeses are tackled afterward. The hardest cheeses, Alpine cheese, cheddar, and Gouda, near the end of the book, are the most complex to make, require a larger quantity of milk, and are best attempted by those with some experience handling smaller and fresher cheeses; however, their aging may actually be easier to manage than smaller, softer rennet cheeses such as Camemberts and washed-rind cheeses—you'll just have to be more patient with them, as they take considerably longer to age!

Chapters 8 through 12 cover rennet-free and lightly renneted cheeses, the most straightforward of cheeses, and the most easily made at home. **Chapter 8: Kefir** looks at the versatile culture in kefir grains, which can be used as a starter for making almost all of the cheeses of this book. **Chapter 9: Yogurt Cheeses** introduces the simplest class of cheese, made by hanging yogurt and allowing its whey to drain. **Chapter 10: Paneer** looks at heat-acid cheeses that are made by heating milk and adding acid to separate its curds.

Chapter 11: Chèvre explores the first rennet cheese of the book, which is more akin to the making of yogurt cheeses. And **Chapter 12: Aged Chèvre Cheeses** looks at ways to take the soft chèvre curd and age it into delicious small surface-ripened cheeses.

Chapter 13: Basic Rennet Curd describes the foundational technique from which most other rennet cheeses evolve. The remaining chapters in the book cover the specifics regarding each class of cheese that evolves from these curds. Learn to make these curds well and you'll have the basic understanding needed to start every other rennet cheese in this book.

Chapters 14 through 21 cover the diverse ways of handling rennet curds made in chapter 13. **Chapter 14: Pasta Filata Cheeses** describes how to transform basic rennet curds into mozzarella and other stretched-curd cheeses. **Chapter 15: Feta** looks into the method of brining as a means of preserving rennet cheeses. **Chapter 16: White-Rinded Cheeses** examines the traditional method of encouraging white fungus to bloom on the rind of a rennet cheese. **Chapter 17: Blue Cheeses** explores how to cultivate the fungus responsible for making cheeses blue, then offers three different methods of making blues. **Chapter 18: Washed-Rind Cheeses** investigates this stinkiest class of cheeses that are made by washing cheeses' rinds with whey as they ripen.

Chapter 19: Alpine Cheeses explores methods of handling basic rennet curds that result in firm and long-lasting cheeses. **Chapter 20: Gouda** investigates the method of washing curds with hot water to firm them, while **Chapter 21: Cheddar** explores the unique way that curds are cheddared to make them into a cheddar cheese.

At the end of it all are two chapters that investigate what to do with the leftover whey and cream that cheesemakers are blessed with. **Chapter 22: Whey Cheeses** looks at the many different ways to handle whey, including making ricotta. And the final chapter, **Chapter 23: Cultured Butter**, explores a method of buttermaking that's surprisingly similar to cheesemaking and makes a surprisingly flavorful butter.

Following the Recipes

Recipes are written for a certain, minimum quantity of milk. They can be easily scaled up, provided you've the appropriate tools and equipment, but they do not scale down well; keeping the temperature of small pots of milk constant can be challenging, as can measuring out very small doses of rennet. Plus, you'll be doing the same work making cheese if you use 1 quart of milk, 1 gallon, or 10 gallons, so why not make more?

Recipes are meant to use good milk—see chapter 2. Milk that is not as good for cheesemaking, because it is from confined animals, or because it is pasteurized and homogenized, or over a week old, will not respond the same way as fresh pastured raw milk, and could result in a failed cheese.

The only two sources of the community of micro-organisms that make natural cheesemaking possible are raw milk and kefir. If you haven't access to raw milk, adding kefir culture to pasteurized milk can help restore the microbial community that is devastated by pasteurization.

Kefir, in fact, is recommended as a starter culture for making just about all of the cheeses in this book, even those made with raw milk. A multipurpose and sustainable cheesemaking culture, kefir can make a simpler and more natural cheesemaking more attainable. I highly recommend finding some kefir grains in your community or online, as they make the methods of this book more approachable. Much more information on sourcing, keeping, and using kefir culture will be provided in chapter 8.

As an alternative to using kefir as a starter, I describe in appendix B, Whey Starters, a method for keeping a whey starter. This method, however, is dependent on raw milk.

Recipes in this book are timed for the use of raw milk from pastured animals; if you use other milks, understand that the curd may not respond in the same way, and that the timings of the recipes may not correlate.

Also, recipes recommend the use of natural calf's rennet, available from cheesemaking supply houses, or made yourself according to the method described in chapter 4. I do not use synthetic or genetically modified rennets in my cheesemaking, so I cannot offer recommendations for their use—from what little experience I have using these rennet substitutes, they change the way the cheeses evolve and may make the recipes difficult to follow.

A Word of Warning

Many of the techniques described in these pages are not commonly practiced. Moreover, it is likely that in most places some of these methods may not even be permissible in the commercial production of cheese. For example, I recommend the use of raw milk for *all* the cheeses in this book, including fresh ones, yet in most jurisdictions the sale of raw milk cheeses aged less than 60 days is illegal. If you wish to sell cheeses made using these techniques, you could lose your production license!

I know that, in my locale, I am not permitted to sell cheeses made in the manner of the book, and this is likely the case in yours as well. But I cannot account for all of the various and shifting regional regulations with regard to raw milk access and cheese production, so I leave it up to you to consult with local food regulations to determine your local legalities. I have essentially written this book as if regulations had no control over the manner in which we make our cheese . . . if only this could be the case!

I personally choose to ignore my local legalities, for I feel they stand in the way of my right to make cheese naturally, and I make my cheese exactly as I please! Fortunately, as a home cheesemaker, there are no restrictions on the methods of your cheesemaking; you are free to make raw milk cheeses as you choose (so long as you can get the raw milk!). However, as a commercial cheesemaker, if you wish to produce cheese in this way and find you're not permitted to, don't just give up hope; consider advocating for your right to practice a natural and traditional cheesemaking.

CHAPTER TWO

Milk

Good milk—naturally raised and minimally processed—is made for cheesemaking. Cheese reflects the quality of the milk that makes it. But not only is the flavor of good milk intensified into the flavor of cheese, good milk helps cheese evolve according to its natural way.

When a young calf drinks its mother's milk, its fourth stomach produces enzymes and acids that curdle the milk into cheese (see chapter 4: Rennet). The semi-solid cheese is more easily held, churned, and digested by the stomach, and the young animal thus receives more nourishment from its milk. The many methods of making cheese mimic this magic that happens inside young animals' stomachs. But if the animals that provide the milk are deprived of the nutrition they need, or if the milk they produce is overprocessed, milk loses its natural ability to become cheese.

Considerations for Good Milk

Milk is most suited for cheesemaking if it comes from grassfed, pastured animals. Regardless of species, milk makes better cheese if the animals are well nourished and eating their most natural foods.

But it's not just the pasturing of animals that makes good milk; the less processing milk has gone through, the better it responds to the cheesemaking process. Milk from pastured, grassfed animals that is obtained by machine milking and that is pasteurized, homogenized, and refrigerated for two weeks will not make cheese as good as what comes from the same milk milked by hand and used while still warm from the udder.

The key considerations to make when sourcing milk for cheesemaking are:

- That the animals providing it are healthy and pasture fed
- That the milk is unpasteurized
- That the milk is unhomogenized
- That the milk is fresh

According to these qualifications, the best milk for making cheese is fresh, unprocessed milk from pastured animals. Conversely, the worst milk for making cheese is any confined animal's milk that is homogenized, pasteurized, and one or more weeks old. (In other words, most of the milk you'll find in the supermarket!) The natural cheesemaking ability of such a milk has been lost, and making cheese with it will present challenges.

Cheesemakers *can* work with overprocessed, industrially produced milk, but they must add back to it what has been lost by the milk's processing and production methods: Bacterial and fungal cultures, enzymes, color, and even calcium are often added to help poor-quality milk make cheese. But despite all these efforts, overprocessed milk will never attain the cheesemaking quality of its raw pastured counterpart.

Don't settle for your average grocery-store-bought milk; understanding the reasons these considerations make a better cheese will help you choose a better milk.

Pasture Feeding

Milk responds to cheesemaking and makes the finest-tasting cheese if it comes from animals that are fed their most natural diet: pasture. Pastured animals nourished on green grass and other herbs produce good milk for cheesemaking, while animals kept in confinement, and fed grain or silage, produce milk that is much less responsive. There is a strong relationship between animal welfare and milk's natural ability to make cheese: The best milk for cheesemaking comes from the best-treated beasts!

Purchasing milk labeled as grassfed or pasture raised is a good bet for ensuring that the milk has come from well-nourished animals, and that the milk will evolve well into cheese. Buying certified organic milk also gives some assurance that animals are pastured; however, the minimum pasturing requirements for organically raised milking animals (that the animals are pastured for 130 days, and that the pasture provides 30 percent of the animals' nourishment) are not necessarily enough to make a better milk for cheesemaking, especially during the fall, winter, and spring when the animals are often kept confined.

Confining milking animals creates a cascade of consequences, not only to the loss of milk's best cheesemaking qualities but also to the ecological footprint of dairy, the health and well-being of the animals, and ultimately the need to pasteurize the milk—to the further detriment of the cheese made from it.

The milk from animals kept in confinement forms weaker curds than pastured animals' milk; milk from cows fed haylage, silage, or grain is deficient in

The best milk for cheesemaking comes from the best-treated animals.

minerals and vitamins that affect curd development. When their animals are off-pasture, cheesemakers find that they cannot make certain cheeses as a result of the milk's changing composition. The making of Alpine cheeses, for example, demands milk from pastured animals, whose curd stands up to the more rigorous Alpine cheesemaking method.

Cheeses produced from the milk of animals kept in confinement are also not as flavorful as those made from animals on pasture. Animals eating their most natural diet produce milk that tastes better, and therefore makes better-tasting cheese. One of the most important reasons that raw milk cheeses taste as good as they do is that the animals that produce their milk are usually nourished on pasture.

Pasturing also makes possible a more ecological cheesemaking. Milking animals kept in confinement have significantly higher environmental footprints than those raised on pasture. From the carbon costs of growing their grain, to the environmental costs of the waste they generate, and the increased use of GM feeds, keeping animals confined doesn't make ecological sense; however, animals raised on pasture provide their own feed and return their wastes to the soil, rebuilding the pastures that sustain them.

For now, the soy and corn and likely the alfalfa present in many total mixed rations (along with many of the added vitamins) fed to confined animals are very likely to be of GM origin (unless the feeds are certified organic or GMO-free). Even the beet pulp fed to confined goats to improve their fiber intake (which they should be getting from their browsing of woody plants) is sourced from predominantly GM sugar beet production, and may have higher residual levels of herbicides (such as glyphosate, the key ingredient in Roundup) that can affect the flora of their digestive tract.

The health of confined milking animals is compromised because of two significant changes to their husbandry: their diet and their living conditions. A grain-based diet overwhelms milking animals' digestive tracts, making them more susceptible to illness; and being kept in closed quarters increases their exposure risk. A life lived on pasture eases these two threats to milking animals' health and leads to an improvement in the microbiological quality of their milk.

Pasture may be as effective as pasteurization at reducing the risks associated with consuming raw milk and raw milk cheeses. Pasturing may thus make raw milk safer to drink, and raw milk cheeses safer to eat, even when they are fresh and not aged for the most often prescribed 60-day minimum. Raw milk from confined animals, however, may be dangerous to drink and to make cheese with. Pasteurization can make such milk safe to drink; however, pasteurizing poorly produced milk only further degrades its cheesemaking ability!

Pasteurization

Pasteurization is a brief cooking that destroys raw milk's native microorganisms and foreign bacteria that find their way into milk as it is handled, as well as any possible pathogens. Pasteurization may make questionable raw milk safe to drink and increase milk's shelf life; it does not, however, result in a better milk for cheesemaking.

Milk can be pasteurized at home by heating it to 143°F (62°C) for 30 minutes stirring all the while, then quickly cooling it in a process known as bulk pasteurization. Commercially, milk is generally pasteurized via a high-temperature, short-time (HTST) treatment at 167°F (75°C) for 15 seconds. Ultra-heat-treated (UHT) milk is flash-heated under pressure to 275°F (135°C) for 2 seconds.

The higher the temperature of the heat treatment the worse the effect on the milk's cheesemaking ability; bulk-pasteurized milk maintains some of its cheesemaking ability, whereas UHT (aka ultrapasteurized) milk does not. Most grocery-store-bought milks are HTST-pasteurized, but an increasing number are UHT: Milk that has very wide distribution needs UHT processing to withstand its longer transit times, and many consumers prefer the longer-lasting milk that results—but again, that longer shelf life translates into less cheesemaking ability.

Thermised milk, increasing in popularity in industrial artisan cheese production, is heated to 145°F

(63°C), but only for 15 seconds. The short cooking time kills the vast majority of raw milk flora along with potential pathogens but leaves the milk's beneficial enzymes intact. Thermised milk maintains some of the raw milk's important cheesemaking qualities but loses the many benefits that come from raw milk's microbiodiversity. Thermised milk cheeses are still legally considered raw, but they won't taste as good as equivalent cheeses made with true raw milk.

All of these methods of heat-treating milk complicate the cheesemaking process, because the native bacteria, fungi, and yeasts in raw milk that are destroyed by pasteurization are necessary for the proper development of cheese. Pasteurized milk is a blank slate devoid of protective microorganisms that is susceptible to spoilage as a result of the growth of unwanted bacteria and fungi. Pasteurized milk must therefore have its microbial community rebuilt through the addition of starter cultures, ripening bacteria, and fungi if it is to be made into cheese.

Pasteurized milk also produces weaker curds when transformed into cheese. Curds made from raw milk develop faster and are stronger, more resilient, and less likely to break when handled than pasteurized milk curds. Cheesemakers often add calcium chloride back to their pasteurized milk in an attempt to restore the mineral balance that strengthens raw milk curds but is lost as a result of pasteurization.

Furthermore, pasteurization degrades the flavor of cheese. The other reason raw milk cheeses are known to have much more complexity than equivalent pasteurized milk cheeses is due to the diversity of raw milk microorganisms and intact enzymes, which creates a more nuanced and flavorful cheese. You'll find a more in-depth exploration of the healthfulness and ethics of raw milk, as well as advice on how to access it and considerations for making cheese with it, at the end of this chapter.

Though good cheese can still be made with pasteurized milk from grassfed cows, if the milk is also homogenized, the combined effects of the pasteurization can challenge a cheesemaker to turn it into good cheese.

Homogenization

Homogenization, a physical processing that stops cream from separating, involves sending milk through a fine-mesh screen under very high pressure. The turbulence caused by the forceful handling causes milk's fat globules to break up into finer particles that no longer rise and separate into cream, remaining instead dissolved homogeneously throughout the milk.

But homogenization doesn't just affect the milk's cream; it also affects milk's cheesemaking ability. With the exception of paneer (see chapter 10), it's just not worth making cheese with homogenized milk, because this most damaging type of processing changes the way the milk reacts with rennet to form into curd. Homogenization results in curds that are weaker and less able to stand up to the rigors of cheesemaking. They fall to pieces, resulting in difficulty handling the cheese, poor cheese yields, and improper cheese evolution: Homogenized milk has such a different nature that many cheesemakers hesitate to call the stuff milk.

The reason for homogenized milk's frailties is that milk's calcium, integral to the development of strong curds, is modified by homogenization, and becomes less bio-available. (That information isn't written on cartons of homogenized milk sold at the grocery store.) Cheesemakers who use homogenized milk must add calcium chloride to their milk to undo the damage; however, despite their best efforts, their curds will never regain their original strength.

Unfortunately, most milk sold in North American supermarkets is homogenized, and therefore not suited for cheesemaking. Not only is "homogenized milk" homogenized, but 2 percent, skimmed milk, and cream are homogenized, too. The only sure way to be certain a milk is unhomogenized is to look for its creamline: Cream rises and separates from unhomogenized milk and leaves a distinct line where the cream and milk meet. Often sold in glass bottles and labeled as "old-fashioned," "standard," or "creamline," unhomogenized milk is making a comeback as consumers demand less processed foods.

If you cannot find an unhomogenized milk locally, call up your local dairies and let them know that you're looking for less processed milk: they may just listen!

Fresh Is Best

Cheese is best when made from milk within a few hours of its milking, for milk is in its best shape when it is fresh from the udder. Fresh milk has intact minerals and a balanced community of bacterial cultures; it's most likely to produce consistent results.

Milk was never meant to see the light of day, and certainly not the light of the refrigerator! Milk, like all other produce, deteriorates as it is cold stored: Refrigeration helps to preserve milk, yes, but the cold temperatures only slow its degradation. Milk is exceptionally well suited to the growth of microorganisms even if kept in cold storage. If milk is refrigerated for too long, it can play host to cold-loving bacteria known as psychrotrophs (Greek for "cold lovers"). Psychrotrophs cause raw milk to thicken and pasteurized milk to spoil even when kept cool.

If you are working with milk (pasteurized or raw) that has been refrigerated for more than one week, it is likely that these unwanted bacteria will have already taken hold: and when that milk is transformed into cheese, the cheese will come to be dominated by these invaders and will develop flaws. Regardless of how good the milk was when it was bottled, if the milk is weeks old, it will have lost its ability to make good cheese.

Common defects that result from using older raw or pasteurized milk are bitter flavors and cheeses that bloat as a result of unwanted strains of fermentative bacteria; I once even had a washed-rind cheese that turned green when I used three-week-old raw milk! Pasteurizing older milk may eliminate the growth of these unwanted microorganisms, but the combined effects of the milk's long refrigeration and pasteurization (a second time if need be) will surely destroy its cheesemaking ability.

For the best possible cheesemaking, don't get your milk in the supermarket! Milk is often already several days old when it arrives in the store and can

Milk, like all other produce, deteriorates as it is cold stored, losing its best cheesemaking qualities.

sit on display for several weeks. Getting milk straight from the udder or—the next best thing—direct from the farmer will ensure the freshest possible milk, and therefore the best evolution of your cheese. If the supermarket is your only option, be careful of what you buy: The First-In-First-Out system of supermarket shelf stocking places the oldest milk in front of the freshest! Check with the dairy department to see when they get their milk delivery, and purchase your milk (being sure to check the "best before" date, which in my mind ought to be a 'milked-on' date) and make cheese soon thereafter. Not sure if your milk is fresh enough? Taste it! Milk should be sweet and clean tasting, not bitter, sour, or rancid—flavors that develop as milk is stored. It's always a good idea to get a sense of the flavor of your milk before setting out to make cheese with it.

Milk Types

There are two other important considerations for a cheesemaker who is choosing milk: Milk can come from many different species and breeds of animals, and it can be either skimmed or whole. These considerations do not affect the cheesemaking ability of milk the same way as those already discussed, but they do still change the cheese that results.

Cow versus Goat versus Sheep versus Buffalo

Though certain species' milks are preferred for making certain cheeses, cows' milk, goats' milk, sheep's milk, and buffalo milk will all respond well to almost every cheesemaking process. Buffalo milk is preferred for mozzarella, and goats' milk for chèvre, but any animals' milk can be used to make any given cheese, and the results will be delicious.

So long as the animals are pastured, any species' milk will form curds that are equally strong. Pastured goats' milk forms similar curds to pastured cows' milk, sheep's milk, or buffalo milk, and all can be transformed into cheese in a similar manner. Cheesemaking guidebooks often recommend adding calcium chloride to goats' milk to help strengthen its curds, but goats with plenty of browse produce milk whose curds are as firm as the finest pastured cows' milk.

Differing milk-solids contents do give certain milks an edge in production over others. Buffalo milk and sheep's milk, for example, contain nearly double the solids of cows' and goats' milk and therefore yield almost twice as much cheese per gallon. Buffalo and sheep also produce milk with higher butterfat content than goats or cows. Buffalo milk is considered the finest for mozzarella making, in part because of its butterfat, while sheep's milk is believed to make the finest blue cheese because the milk is so rich.

Different animals' milks will change the color of a cheese. Goats' milk will give a cheese whiter flesh, as will buffalo milk and sheep's milk; cows' milk, however, has a creamier hue, particularly if the animals are pastured and have been eating fresh green grass. Animals that give cream-colored milk pass the carotene pigment (which is also a precursor to vitamin A) in green grass through their milk, whereas animals that give white milk transform carotene into the colorless vitamin A.

Each breed of cow, goat, sheep, and buffalo also yields a milk of distinct quality, and each different breed's milk will therefore produce a distinct cheese. Milk from a Jersey cow, for example, has a higher butterfat and a creamier color than milk from a Holstein; a Jersey milk cheese will therefore have a higher butterfat and creamier color than a Holstein milk cheese.

Whole Milk versus Skimmed Milk versus Cream

Whole milk and skimmed milk will both respond well to cheesemaking so long as they are minimally processed. Cheeses generally turn out better, though, if made from whole milk: Its higher butterfat gives cheeses a creaminess that skimmed milk cheeses lack. Skimmed milk cheeses tend to be more rubbery, less flavorful, and generally not as appealing.

Using whole milk also results in a greater yield: Whole milk makes almost twice as much cheese as the same volume of skimmed milk. The higher butterfat content in whole milk means more milk solids, which means more cheese!

Some rennet cheeses are even made with an additional dose of cream. Double- or triple-cream Bries have cream added to the milk to make them more luscious. However, the cream is added back in the form of crème fraîche (see chapter 8, Kefir) so that it can be incorporated into the curds, because separated cream does not respond to cheesemaking the same way as milk.

If unhomogenized milk has sat refrigerated for several days, its cream will separate to the point that it will not incorporate into the curds. Separated cream, which does not contain the casein protein that forms curds, can be left behind by the cheesemaking process and may remain in the whey that is strained from the curds. Most artisanal cheesemakers therefore use their milk the same day it is milked in part so that the cream doesn't have a chance to rise.

Regularly stirring milk as it is stored can help ensure that its cream does not separate. If your cream has already separated, skim it before you make cheese. And if you find cream left behind in the whey, it, too, can be skimmed; such whey cream is used to make whey butter—see chapter 23 for more information.

Raw Milk and Cheesemaking

Raw milk is not inherently dangerous but rather a source of all that is nourishing and healthful; it is the way raw milk is often produced that makes it dangerous!

Do not trust just any raw milk to be good for cheesemaking. If it was produced using standard industrial dairying practices, raw milk can be dangerous to drink and should not be made into cheese even if it is aged the requisite 60-day minimum required for raw milk cheeses. But raw milk, produced to the high standards of quality that make it safe to drink, will be safe to make into cheese, even if the cheese is consumed when young and fresh.

Raw milk can be safe if it comes from healthy, well-raised animals. Access to fresh air and fresh pasture, conscientious milking practices, clean housing, and regular testing of the herd and their milk can all help to ensure that raw milk is safe to drink, safe to sell, and safe for making cheese.

I cannot offer much more advice on practices that can improve the quality of raw milk; this is a cheesemaking guidebook, after all, and not a raw milk production manual. For more information on good raw milk protocol, consult the sources at the end of the book.

The quality of milk as observed through the lens of cheesemaking no doubt reflects upon the quality of milk with respect to human nutrition. If the making of cheese mimics the processes of milk digestion, then a milk that responds poorly to cheesemaking may very well respond poorly to digestion . . . for it likely does not contain the minerals and vitamins present in good milk. Raw milk from pastured animals responds best to cheesemaking—and this, I believe, is the greatest evidence of any milk's healthfulness.

Concern about the safety of raw milk has led to restrictions on its sale in many jurisdictions around the world. And understandably so . . . for raw milk can be a serious threat to public health. To contextualize the prevalence of pasteurization, and to understand the reasons it is mandated for the treatment of milk, a short history lesson is in order.

A Short History of Pasteurization

Social and economic developments that caused rapid urbanization in North America at the end of the 19th century caused enormous changes to dairying practices, resulting in health crises that ultimately led to legislation requiring milk's pasteurization in most jurisdictions.

As urban populations grew, the demand for fresh milk began being filled by urban cowsheds that could provide milk without a lengthy trip to town, during which time it was prone to souring. However, the conditions at many of these urban cowsheds (also known as swill dairies) were far from ideal: Dairy cows were kept in confined quarters, without access to fresh air, clean water, or pasture, and the animals were fed inferior feeds, including spent grains from breweries and distilleries. Conditions at such dairies resulted in unhealthy livestock and caused their milk to be a source of serious illness.

Raw milk from swill dairies was recognized to be dangerous, particularly among infants and the elderly. And it is here that pasteurization came into the picture. Originating as a method of controlling unwanted fermentation in winemaking, pasteurization's first use in dairying was to sterilize milk to stop it from souring as a result of spontaneous fermentation on its long trip from country to city. However, it was soon realized that pasteurization didn't just stop raw milk from souring by killing milk's native bacterial cultures but also killed off the disease-causing agents present in poor-quality raw milk. Illnesses like tuberculosis and brucellosis that were linked to consumption of tainted raw milk seemed to be kept under control with pasteurization.

As illnesses arising from poor dairy practices spread, two schools of thought emerged about what to do about "the raw milk problem." One camp believed that milk should only be sold to the public if it came from certified dairies that kept their animals

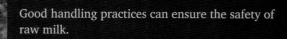
Good handling practices can ensure the safety of raw milk.

on pasture and produced milk up to a very high standard that was safe to drink raw. The other camp believed that farmers could not be trusted to produce milk safely, that milk was commonly adulterated as it was handled, and that the only way to make milk safe for consumption was to "Pasteur-ize" it!

Larger dairies, fearful that their operations would not be able to profitably produce milk up to the standard of certified dairies, preferred the path of pasteurization. Small-scale dairy farmers preferred the option of certification; their farms were likely already producing up to that standard. Guess which side won?

The Freedom to Choose Raw Milk

The practice of pasteurizing milk thus arose to combat health crises resulting from poor dairy practices more than a hundred years ago. Today we live in somewhat more enlightened times, and yet we are stuck with regulations put in place a century ago that favor the industrial approach to dairying. We now understand the conditions that cause raw milk illnesses to arise; we have technology that can test animals and milk for evidence of illness; and we are capable of putting in place certified production standards that confirm the quality of raw milk. And yet consumer access to safe and well-produced raw milk is restricted by outdated pasteurization regulations that supposedly protect the safety of consumers, but really protect only the interests of large-scale milk producers.

If raw milk is not legally available in your locale, don't let that stop you from sourcing it. Just because it is illegal for a farmer to sell raw milk does not necessarily mean that it is wrong for that farmer to do so. Bans on raw milk sales are supposedly put in place to protect consumers from dangerous raw milk; but if raw milk comes from animals that are well raised, and is produced with conscientious milking practices, it will not be dangerous and need not be kept out of reach of the public. Perhaps there are other reasons raw milk doesn't flow so freely!

Big Dairy continues to stand by pasteurization because the technology allows its farmers to raise animals in confinement and still safely sell their milk to the public. Raw, industrially produced milk can have higher levels of unwanted bacteria and pathogens in it before it is pasteurized, because pasteurization eliminates the risks from consuming poorer-quality milk. But for instance, the standards for somatic cell counts (SCC), a count of white blood cells in milk and an indicator of the health of milking animals, are considerably stricter for certified raw milk than for milk destined to be pasteurized.

The criminalization of raw milk sales results in the criminalization of sustainable and small-scale dairy farming. Forced pasteurization has destroyed countless small dairy farms that once sold their milk locally and could not afford the cost of the new technology that was required to process and sell their milk. And these regulations continue to restrict small-scale dairy farmers who wish to produce good raw milk to sell on the local market. Farmers need to be trusted to produce raw milk safely so that it can be made available to those who desire it; and farmers should be rewarded for producing higher-quality raw milk with the right to sell that milk, and for a higher price.

We as consumers have the right to choose food that is produced up to standards of quality that we believe in. We choose to purchase food that has been produced organically, and to kosher or halal standards, and we should have the same right to choose raw milk that is safely produced to a higher standard than industrially produced milk.

We have easier access to many products that are considerably more dangerous than good raw milk like alcohol, firearms, cigarettes, high-fructose corn syrup, and motor vehicles; and with the Centers for Disease Control (CDC) reporting but *two* deaths in the United States traced to raw milk consumption between 1999 and 2011, the risks of raw milk are comparatively nil! Perhaps we should trust consumers to make an informed choice to purchase raw milk and raw milk cheeses, even if they do carry a slight risk.

Restrictions on raw milk severely constrain the culture of cheesemaking, create an underground or "black" market for raw dairy, limit sustainable dairying practices, and put farmers' livelihoods at risk. I believe that many advantages would come if consumers were allowed greater access to good raw milk.

Sourcing Good Raw Milk

Good raw milk makes beautiful cheese. But most home cheesemakers will have difficulty sourcing it because of regulations controlling its sale. Though it is not expressly illegal for consumers to purchase raw milk, there are generally severe restrictions on its sale, regardless of its quality.

Depending on the jurisdiction, the legalities of raw milk vary widely. Raw milk sales can be outright forbidden, or only permitted at the farm gate. Some states, such as Oregon, only forbid public sales of raw cows' milk but permit the sales of raw goats' milk (evidence to me that the issue is more about politics than public health). And in the most progressive locales, laws permit consumer access to good raw milk from certified producers who raise their animals according to the highest standards of production, and who sell their milk at farmers markets or even in grocery stores.

But in most places home cheesemakers looking for the milk most suited to cheesemaking must either get their own milking animals or find a farmer willing to break the law to sell them contraband raw milk. Where I live, in Canada, raw milk sales are forbidden in every province, and the home cheesemaker's only legal (though hardly sustainable) option to get the best milk for cheesemaking is to cross the border and bring back raw milk from American states that allow direct sales at the farm gate. Though we are legally permitted to bring back a certain number of gallons per trip, we still get the thrill of rumrunners trafficking an illicit substance!

Many farmers who wish to make good raw milk more accessible to the public have found ways around bans on raw milk sales. For example, the illicit act of selling raw milk can sometimes be circumvented through the practice of cow-sharing. Cow shares (or goat shares or sheep shares) operate on the legal premise that if a person owns a milking animal, they are entitled to her raw milk; and that if a person owns a share of an animal, they are therefore entitled to a share of her milk. Farmers who operate cow shares sell shares of their milking animals to their clients, who pay for the upkeep of the animal in return for a share of the raw milk she produces. Cow-share operations, equivalent to CSA produce shares, form stronger relationships between farmers and consumers and promote safe and sustainable raw dairy practices.

Another option is to offer your cheesemaking services to farmers in exchange for their milk. Dairy farmers can loan cheesemakers their milk, which they transform into cheese and return to the farmers, retaining a certain share for themselves.

If you don't know where to start your search for raw milk, ask around (discreetly) at your local farmers market, or become a part of online communities that link consumers with raw milk farmers who are willing to break regulations to make good milk available to the public. There's good raw milk out there, and lots of good farmers who are willing to share.

The moral of raw milk is to make connections with your local farmers, and support their efforts to raise good-quality raw milk, even if its sale is illegal. Raw milk should come from farmers whose concern for the health of their cows and the quality of their milk is foremost. It is a very good idea to know that the farmer is conscientious and understands good raw milk dairying practices, and direct sales offer the greatest assurance of the quality of raw milk next to certification.

Considerations for Making Cheese with Good Raw Milk

If you use raw milk in cheesemaking, there are some important things to know about its nature. Because raw milk still has its native community of microorganisms, because it still has its endemic enzymes, and because it is more minerally balanced than other milks, it behaves better than pasteurized milk when made into cheese. The methods of making cheese with it are therefore simpler and more intuitive, and more likely to produce good cheese. And a cheesemaking recipe that calls for raw milk may not work for pasteurized milk, and vice versa.

Raw milk does not need to have starter cultures added. Several cheeses in this book call for only raw milk's native cultures: Such cheeses cannot be made in the same manner using pasteurized milk. Whey from a batch of raw milk cheese can also be saved for making another batch of cheese with excellent results, because the beneficial raw milk community of microorganisms is cultivated and nourished by the act of making cheese. In the same way that a sourdough starter is nourished by feeding it wholemeal flour, the bacterial community in whey used as cheesemaking starter is nourished by regularly making cheese. Pasteurized milk cheeses, however, should not have their whey saved for making another batch of cheese, because the milk does not contain the necessary balance of microorganisms that promote the development of a healthy microbial ecology. See appendix B, Whey Starters, or more information on keeping a whey starter.

Raw milk does not need to have ripening cultures added to it. The diverse community of bacterial cultures in raw milk provides all the ripening cultures a cheese needs. Raw milk cheeses will even grow healthy white fungal coats without the need to add freeze-dried fungal spores to the milk. The *Geotrichum* fungus naturally present in raw milk establishes healthy white rinds on raw milk cheeses if handled in the appropriate manner; pasteurized milk cheeses, however, will not grow *Geotrichum* coats unless the culture is intentionally added.

Good raw milk evolves faster into cheese. Raw milk curds firm up and form into cheese faster than those formed from overprocessed milk. The cheesemaking process often takes one-third less time if raw milk is used in place of overprocessed milk. The cheese that results is easier to handle and withstands the handling that it receives; poorer-quality milks make curds that are softer, more delicate, and more likely to break if overhandled.

If you only have access to pasteurized milk, using kefir as a starter can help to restore some of raw milk's biodiversity and can help it gain much of what was lost through pasteurization. Unfortunately, kefir doesn't undo the damage wrought by homogenization, or keeping animals confined (more information on maintaining a kefir culture is provided in chapter 8).

CHAPTER THREE

Culture:
The Ecology of Cheese

Like a forest, cheese is an entity with a complex ecology. Countless species of bacteria, yeast, and fungi—sometimes even insects—call cheese home. These diverse species interrelate, compete, and cooperate within the flesh of a cheese in fascinating ways—ways that contribute to the incredible diversity and breadth of flavors of the many types of fresh and aged cheeses.

Empty Cheese Ecologies

It is because of the myriad forms of life living in cheese that cheeses can be aged. A lifeless cheese like paneer, made by boiling milk and adding vinegar, cannot be aged because it lacks the vital bacterial cultures that help cheeses mature. Paneer is a blank slate, an empty ecology containing an abundance of nutrients that wild bacteria, yeasts, and fungi will prey upon; such a cheese will age in unpredictable and possibly dangerous ways.

Pasteurized milk presents another empty cheese ecology. The heat treatment of pasteurization destroys bacteria, both beneficial and pathogenic. The milk that remains is lifeless, with abundant nutrients to support the growth of wild bacteria, fungi, and yeasts that find their way into the empty milk. As pasteurized milk sits in the refrigerator, the low temperatures do not stop these wild

bacteria from doubling their numbers every day. If cheese is made with several-weeks-old pasteurized milk, it is recommended to pasteurize it *again* before cheesemaking to ensure that the wild microorganisms do not come to dominate the development of the cheese.

Because of its empty ecology, cheesemakers must add cheesemaking cultures to their milk if they make cheese with pasteurized milk. However, the freeze-dried bacterial cultures that most cheesemakers use, known as direct vat inoculants (DVIs), do not complete the ecology.

When cheesemakers add DVIs to their pasteurized milk, they are adding individual prepackaged bacterial species into a cheese, not the palette of microorganisms that a cheese truly needs. These strains of bacteria are not stable communities of microorganisms and leave niches open to opportunistic and possibly pathogenic bacteria. Much as in industrial silviculture, where a single species of planted tree cannot create a healthy and dynamic forest ecology, freeze-dried monocultured DVIs do not create a healthy and dynamic cheese ecology.

The Ecology of Raw Milk

The basis of a healthy cheese ecology is provided by raw milk.

The *Geotrichum candidum* fungus of raw milk can help to establish white rinds on aged cheeses.

One of the most significant components of raw milk is its diverse population of beneficial bacterial cultures. The many species of microorganisms present in raw milk help to establish healthy gut flora and strong immune systems in the infant cow or goat or sheep (or human) that drinks it. These diverse cultures also help to make and ripen raw milk cheeses (see appendix E for a list of beneficial cheesemaking microorganisms commonly found in raw milk).

Science is beginning to recognize that raw milk contains healthful probiotic cultures that are an integral component of the milk. It was originally believed that bacteria found their way into raw milk only through environmental contamination, but scientists have recently found that pathways exist within mammals for the growth and propagation of bacterial cultures that are beneficial to young mammals' health. (Infant formula manufacturers are lately clueing in to the importance of probiotics in milk and have begun to add single-strain monocultures of beneficial bacteria to their products, which, still, can never offer the same health benefits to infants as biodiverse breast milk.)

These beneficial raw milk cultures help infant cows, goats, and humans in numerous ways: Gut development, digestion, immune system development, dental health, and resistance to pathogens are all aided by cultures passed from mother to infant. Raw milk cultures, which differ only slightly from cow to goat to human, include relatives of the DVI cultures that are used by contemporary cheesemakers to make their cheeses (see appendix E for a comparison). These raw-milk microorganisms are the original cheesemaking cultures.

In all raw whole foods, the food plays host to beneficial bacteria that are particularly suited to devouring it. These native cultures also help to transform the

The natural yeasts that bloom on grape skins spontaneously ferment its juice into wine.

basic foodstuff into traditionally fermented foods: Cabbage contains all the bacteria it needs to become sauerkraut, wheat has all the bacteria and yeasts it needs to become bread (or beer), and grapes have all the culture they need to become wine. Milk is no exception: The native biodiversity of raw milk provides microorganisms that help infants digest their mothers' milk (and cause the milk to decay if it is spilt); these microorganisms are all that that milk needs to become the many different styles of cheese.

Early cheesemakers relied on these raw milk cultures to make their cheeses. The diverse cultures present in their raw milk helped to fill the diverse ecological niches within different cheeses and contributed the breadth of flavors that raw milk cheeses are known for. When milk is pasteurized prior to cheesemaking, these beneficial bacteria are eliminated, and no amount of DVI cultures can fill in for all of the diverse cultures that give life and flavor to a cheese.

The Life and Death of Cheeses

Cheeses made with biodiverse raw milk and tended to in a cheese cave will evolve in a predictable, degradable way. Forms of life succeed one another in cheeses as they age: The cultures that dominate a cheese change over time, in successions similar to the changing plants in evolving forest ecosystems. Within cheeses, cultures cooperate and compete and pave the way for others to succeed them. Starter bacterial cultures give way to ripening bacteria and yeasts, which give way to fungal cultures, and sometimes even insects, as the cheeses age.

Cheeses are started with lactic acid bacteria, indigenous to raw milk, which feed upon the milk's lactose sugars when the milk is first warmed. The *Lactobacillus* bacteria ferment the lactose to lactic acid and increase the acidity of the milk, allowing the milk proteins to reconfigure and coagulate into curd with the rennet enzyme that is added to the milk (more on rennet coagulation in chapter 4).

As the curds coagulate, and are then cut and stirred, the *Lactobacillus* cultures continue to acidify the milk, making the curds even firmer as the increased acidity pushes their whey out. When the curds are strained from the pot, the acidification continues on. And as the curds coalesce into cheese in their forms, they get more acidic still. Pasta filata cheeses such as mozzarella are allowed to continue acidifying until the point at which their acidity allows the curds to be stretched when submerged in very hot water.

Salting cheeses slows down that acidification. As the salt pulls moisture out of the cheese, its bacterial activity is significantly reduced. Most cultures—bacterial, fungal, and yeast—are hampered by the reduction in moisture and increased salinity within the cheese; however, certain cultures are resistant to the changing conditions, and these continue to affect the ripening cheeses in beneficial ways. Cheeses that are not effectively salted can fall prey to cultures that degrade cheeses in ways that do not improve their flavor (more on salting in chapter 5). After salting, cheeses are ready to be aged, and as cheeses are placed to age in their caves, many complex changes occur on an ecological level.

Below the rinds, a new complex of bacteria succeeds the *Lactobacillus* cultures that started the cheeses. Previously dormant in the milk, these anaerobic (meaning "surviving without oxygen") ripening cultures feed upon the products left behind by the starter bacteria, as well as the dead starter bacteria themselves. As these bacteria do not need air to thrive, they continue to develop and age the cheeses from within. These successions of cultures produce enzymes that help to soften up the paste of the cheese.

Colonizing species of yeasts de-acidify the surfaces of freshly made cheeses. These yeasts, present in raw milk, feed upon the sugars on the surface of the cheese and produce alkaline wastes that neutralize their environment. As they neutralize the surface of the cheese, the yeasts make the rinds more hospitable for the growth of fungal cultures that prefer less acidic conditions.

Fungus, present as spores in the milk, will grow when it finds itself in favorable conditions upon the rind of a cheese. As the fungal spores germinate, they send out their mycelium, a network of fine rootlike filaments called hyphae, and begin to feed upon the cheese. Fungal cultures begin to show their fuzzy growth upon the surface of a cheese within a week of its making.

Because they need air to thrive, fungal cultures affect the cheeses from their rinds toward their centers, a characteristic that can be witnessed in the interior of an aging Crottin, whose ripening begins at the rind and continues toward the center as it ages. Upon the surface of the cheese, the fungal cultures begin to mature, producing their microscopic fruiting bodies that shed spores into the air. A ripe blue cheese shows its true fungal cultures in its colors; when it ripens, its mycelium produces fungal spores that color the cheese blue!

Species of insects may also find their way into a cheese. Tiny spiderlike creatures known as cheese mites, residents of many cheese caves, feed on the fungal growth on the surface of the cheese. As they tunnel through the flesh of the cheese, they allow air to access deeper within and thus encourage even more fungal growth, which they then feed on. The mites that feed on rinds define some cheeses, such as Mimolette: These mites give the cheese its characteristic surface textures and flavors. Most cheesemakers, however, keep their caves scrupulously clean to limit the proliferation of mites.

Fungal cultures eventually degrade the cheeses they feed upon. Known as decomposing organisms, the fungal cultures that make their way into cheeses slowly degrade them. As the fungus runs out of food within the cheese, it drowns in its own wastes; a flood of ammonia, the scent of decay, signals the cheese's ultimate demise. Some people appreciate the strength of character of overripened, ammonia-rich cheeses, but most would throw the dying cheese onto the compost heap.

Changing the Conditions Changes the Ecology

Cheeses are escorted by their maker along a certain pathway toward decay. Each and every aged cheese that we enjoy is made in a particular way that defines its ecological conditions and encourages it to degrade in a well-defined way.

The ecological conditions that are responsible for aging cheeses can be controlled by the way in which the cheeses are made and handled, the conditions of their aging environment, and, to a lesser extent, the cultures that are added to a cheese: The addition of certain cultures does indeed have an influence on a developing cheese, but most of those cultures have their origins in raw milk, and what proves to be much more important in the making of cheese is cultivating the conditions that the right cultures need to thrive.

Different handling techniques condition cheeses to age in different ways. Depending on how the curds are made, the conditions of the cheese will change dramatically: Softer cheeses will age more quickly because their high moisture content encourages bacterial and fungal development, while harder cheeses age more slowly as their lower moisture content slows bacterial and fungal growth.

Cheeses with a flattened shape, such as Camembert and certain washed-rind cheeses, ripen primarily from the surface because of their shape. Cheeses that are made larger are less influenced by fungal growth on the surface and tend to age from the interior. However, piercing a large cheese results in air getting to its interior, which encourages fungal growth on the inside of a cheese where it would not normally exist—a technique used in the making of many blue-veined cheeses.

Different aging conditions encourage cheeses to age in certain directions: If a cheese is kept in more humid aging conditions, fungal cultures may come to dominate, while cheeses that are kept in drier caves will be less influenced by fungi. Cheeses that are exposed to moving air will ripen differently from those kept in closed quarters. A cheese that isn't turned, such as the pyramid-shaped Valençay, will age differently on its top than on its bottom: The top of the cheese will have access to air and will ripen fungally, while the bottom of the cheese will not breathe and will ripen bacterially.

Different cultures can coexist in an aging cave as long as each cheese gets the treatment and respect it deserves.

Changing the conditions changes the ecology: Washing cheese rinds with salty whey keeps fungal growth at bay but results in the growth of bacteria and fungi that give beautiful pink and orange hues.

Whether or not fungal cultures are intentionally added to a cheese, fungus will find its way in, and if a cheesemaker does not intervene, those fungal cultures will define the ripening of a cheese. Of course, not every cheesemaker wants fungus to run rampant over their cheeses, and not every cheese tastes the way it is intended to if fungus is allowed to develop upon it. Indeed, many cheesemaking styles were developed because of the methods that cheesemakers used to keep their rinds free from unwanted fungal growth.

The techniques of brushing, oiling, waxing, bandaging, and smearing rinds, and keeping cheeses submerged in brine, each of which helps to define certain cheeses, all exist to keep fungus in check. Every one of these techniques stops the growing fungus by physically setting back fungal growth or by restricting exposure of the cheese's surface to air, so essential for fungal development. Many Alpine cheeses have their rinds brushed regularly to stop fungus (as well as cheese mites) from growing on their surfaces. Rubbing cheeses' rinds with oil, as is the practice with Parmigiano Reggiano, knocks back fungal growth but also coats the rinds with fungus-resistant oil. Waxing rinds, as is practiced with many cheeses, such as Gouda, keeps air off the surface of the cheese, stopping fungal development entirely. Wrapped cheeses, such as traditional clothbound cheddar, are first wrapped with fine cheesecloth, then slathered with fat, which keeps air off just like waxing does. Aging cheeses by submerging them in a light brine or oil, as is common in the making of feta, eliminates the fungal threat entirely.

Rind washing, also known as smear ripening, is an aging technique unique among the diverse methods of keeping fungus in check. This method of regularly washing cheeses' rinds with a salty brine encourages the development of a new ecology on the surface of the cheese that influences it in fascinating ways. As a result of the regular smearing, these cheeses are kept free of fungus. But because the surface of the cheese is kept consistently wet and salty, the ecological void left behind by the eradicated fungal cultures becomes filled with a new complex of ripening yeasts and bacteria that thrive in these conditions and contribute to these smear-ripened cheeses' bright orange color and stinky aromas.

Traditional Cheesemaking Culture(s)

Of all the cheese styles that exist, none was created in our post-Pasteur era of cultural cognizance. Given the success of ancient cheesemaking practices in developing nearly every cheese we know and love, it's almost as if the very idea of bacterial cultures is a modern lens through which traditional cheesemaking methods are observed.

Cultural practices defined the first cheesemaking cultures. Early cheesemakers had no idea what bacteria were, that they were present in their milk, or that they played a major role in the creation and the aging of their cheeses; nonetheless, cheesemakers created the ideal conditions for these beneficial bacteria to thrive. Their cultural practices—the particular methods they used to handle their milk and their cheeses—encouraged the growth of specific bacterial cultures that defined their cheeses. And diverse cultural practices from traditional cheesemaking cultures around the world created diverse ecological conditions that gave birth to thousands of distinct varieties of cheese.

Traditional Starter Cultures

The original cheesemaking starter culture was the sloppy milk bucket. Cheesemakers never washed out their milking buckets; thus, native raw milk bacteria thrived in the bucket, and all the raw milk that came into it was inoculated with large numbers of beneficial bacterial cultures that improved the ecological conditions for making cheese. Few contemporary cheesemakers would never wash out their buckets, but for thousands of years traditional cheesemaking culture thrived on what would be considered unsanitary or questionable practices today.

Early cheesemakers evolved methods of keeping starter cultures through the unsavory-sounding but

THE ART OF NATURAL CHEESEMAKING

effective practice of backslopping: saving a small amount of a previous cheesemaking's whey as a "mother" to start the next batch of cheese in a practice similar to sourdough bread baking. Through backslopping, cheesemakers kept their cheesemaking cultures alive just by making cheese regularly, nourishing and invigorating their cultures with every batch they made. Many traditional European cheeses such as Parmigiano Reggiano are still made this way today.

Each region would produce a cheese that was distinct, due in part to unique bacterial starter cultures that are summoned from the native raw milk cultures of that region. The bacteriological profile of raw milk is indeed affected by many factors, including the species of animal, be it cow, sheep, or goat; the particular breed of animal, which affects various qualities of their milk; and the profile of plants that are included in the animals' diets. And each region had its own particular species, breeds, and feeds that defined the cultures that in turn helped to define their cheeses.

Traditional Fungal Cultures

Fungal ripening cultures were also cultivated through traditional cheesemakers' cultural practices. Certain cheesemaking traditions, such as using raw milk, would have helped ensure healthy fungal growth on cheeses. In addition, leaving cheeses to age in cool and humid caves exposed them to wild fungi from the earthen walls that would have colonized their rinds. Historically, cheesemakers also found ways of propagating different strains of fungal cultures that helped age their cheeses.

Early cheesemakers inoculated their cheeses with specific fungal cultures and created the conditions that most favored their growth long before fungal cultures were discovered. *Penicillium roqueforti*, the famous fungus that makes cheese blue, was propagated by traditional cheesemakers on sourdough bread, upon which this famous fungus also grows. Cheesemakers then added the moldy bread to their milk to encourage their cheeses to turn blue!

The fungal cultures that give cheeses their white rinds were also cultivated by particular cheesemaking practices that favored their growth. *Geotrichum candidum*, a fungal culture native to raw milk, colonizes the rinds of traditional raw milk cheeses handled in the right way: White-rinded cheeses such as Camembert and Crottin were originally ripened with the aid of only these raw milk cultures. The particular conditions in which the cheeses were made favored the growth of *Geotrichum* and other native raw milk fungal cultures that result in the development of beautiful white rinds.

The cheese-ripening environment also served as a rich source for fungal cultures that ripened cave-aged cheeses. The walls of cheese caves harbored diverse fungal cultures native to raw milk that thrived upon the cheeses aged there, as well as wild soil fungi present in the earthen walls. These fungal cultures decompose organic matter in soil and help recycle nutrients, which is exactly what they'd do when they found their way to the surfaces of the cheeses in the cave.

Freeze-Dried Cheesemaking

The cheesemaking practiced today in North America is dependent upon the use of laboratory-raised, freeze-dried strains of bacteria and fungi. Paralleling the monopolies of seed companies, these packaged cultures are produced by a decreasing number of culture houses that are being consolidated and bought out by some of the better-known agribusiness and biotech giants. Cheesemakers who use these cultures may be unknowingly supporting corporations that are set on developing not only genetically modified seeds but also genetically modified cheesemaking cultures.

Freeze-dried cultures, originally isolated from traditional mother cultures from across Europe, are advertised as being simpler to keep, better able to provide consistency in the final product, and equal to traditional cultures when it comes to flavor development. Beginning in the 1970s, "freeze-dried cheesemaking" became the norm, and the practice of keeping mother cultures fell from favor. Today few cheesemakers keep their own cultures. And home cheesemakers who wish to take up the art of cheesemaking follow suit.

The stated advantage of freeze-dried DVI cultures is their ease of use. Certainly the practice of keeping cultures is more involved than the use of DVIs, but given the disadvantages and shortcomings of DVIs, the case for keeping mother cultures grows.

Disadvantages of DVIs

Freeze-dried cultures promote unsustainable cheesemaking practices.

Cheesemakers cannot reculture their DVIs. They are unstable collections of bacteria that do not exist in nature, whose profiles change over time, and whose performance and quality will decline if reused. Much like the unpropagatable hybrid seeds that many farmers use, DVIs must be purchased anew for every batch of cheese. Cheesemakers, therefore, become dependent upon purchased packaged cultures. Even if it were possible, reculturing the proprietary blends of often trademarked cultures may not even be permitted!

DVIs are not adaptable to different styles of cheesemaking. The two or three strains of bacteria in each package are selected for making cheeses in a certain range of conditions and cannot be used for making cheese of a different style. Cheesemakers are thus instructed to buy different cultures for the different cheeses that they make.

Cheesemaking supply companies sell many strains of starter bacteria, some of which are mesophilic (suitable for making cheese at lower temperatures) and others thermophilic (suitable for making cheese at a higher temperature). To make a broad spectrum of cheeses, supply companies recommend that cheesemakers buy half a dozen different DVIs: Those who attempt to make a diversity of cheeses such as chèvre, blue cheese, Camembert, Parmesan, feta, and Gouda are instructed to buy a package of cheesemaking culture for each type!

DVIs require sterile conditions to thrive. Because they are unstable, arbitrarily selected strains of bacteria and not stable communities of microorganisms, cheesemakers are forced to adopt sanitized working practices to create the conditions that these laboratory-raised cultures need to thrive. Sterilization of the milk, the tools, and the hands of the cheesemaker are essential to ensure that these sensitive cultures do not fall prey to unwanted wild cultures that might derail a cheese.

Bacteriophages, viruses that can destroy entire populations of DVI cultures, are of great concern to cheesemakers who use freeze-dried cultures. As a result of the threat of bacteriophages to monocultured DVIs, cheesemakers can never be certain that their cultures are active and must constantly monitor the acidity of their vats to be sure. Cheesemakers are encouraged to switch up their starter cultures regularly to stay one step ahead of the phages.

The cultures in DVIs do not persist through a cheese's ripening stages. The bacterial cultures present in aged DVI cheeses are often not the DVI cultures introduced by cheesemakers into their vats. As cheeses age, their ecologies shift due to changing conditions, and DVIs do not include the array of cultures necessary to promote healthy cheese aging. Waves of successive ripening bacteria from the wild find their way into DVI-started cheese and can take over the ripening in possibly unpredictable ways.

The costs of DVIs can quickly add up, especially for home cheesemakers, who pay more per dose than commercial cheesemakers that buy their cultures in bulk. And DVIs do not last forever: Many home cheesemakers, who only make cheese occasionally, often find that their cultures have expired before the packages are finished.

By purchasing DVIs and supporting culture corporations, we lose control of the culture of our cheese. Many culture houses have proven their devotion to the development of GM technologies. Genetically modified yeasts are already being used in the production of many other fermented foods such as bread, beer, and wine. Given their speedy approval by governments, it is entirely possible that GM starter cultures could someday become available for North American cheese production.

We need to stop thinking of the cultural regimes in cheese as individual strains of bacteria or fungi. It is the reductionist industrial approach to cheesemaking that has singled out what are believed to be the important players in an evolving cheese; the minor players have been cast aside and forgotten. Yet it's the combination of the important players and their supporting microorganisms that make for a healthy cheesemaking. For it is not an individual species that makes a cheese, but a community of microorganisms, that is as yet too complex to examine or model, that consists of many species of bacteria, yeast, and fungi, all living together and evolving together.

Even I fall into the trap of naming the individual species responsible for the particular characteristics of certain classes of cheese. It's time we shed ourselves of this lens through which we perceive our cheeses and take a more ecologically sound, community-minded approach to the culture of cheese.

The Black Sheep Approach to Culture

In the decade that I have been practicing cheesemaking, I've never once used direct vat inoculants. From the moment I started making cheese I knew that the culture of cheesemaking shouldn't come from a package.

Through experimentation and the study of traditional cheesemaking practices, I've compiled methods of cultivating all the cultures a cheesemaker might need without having to resort to buying *any* freeze-dried cultures. Among the "counter-cultures" (that's how I refer to the cultures I keep on my counter) I keep for cheesemaking are kefir for starting cheeses, *Geotrichum candidum* fungus for white-rinded cheeses, *Penicillium roqueforti* fungus for blue cheeses, and *Brevibacterium linens* bacteria for smear-ripened cheeses.

Kefir as Starter Culture

Kefir, to be discussed in greater detail in chapter 8, is a versatile and sustainable cheesemaking culture.

A traditional dairy ferment from Central Asia, kefir contains a stable and diverse community of bacteria, yeast, and fungi that form colonies known as kefir grains. As easily kept at home as a sourdough starter, kefir is incomparably useful for cheesemaking.

Kefir can be used as a stand-in for any freeze-dried DVI starter: Its microbial diversity allows it to adapt to any style of cheesemaking (see appendix E for a comparison of kefir's cultures with packaged DVI cultures). Chèvre, usually started with freeze-dried mesophilic bacteria, can be started with kefir, which naturally contains mesophilic bacteria; Alpine cheeses, usually started with freeze-dried thermophilic bacteria, can also be started with kefir, which plays host to thermophilic bacteria as well. Interestingly, kefir contains beneficial fungal cultures that can also help to age many types of fungally ripened cheese.

The biodiversity provided by kefir also offers solutions to many of the problems associated with DVI-cultured cheeses, such as bacteriophage threats, changing bacteriological profiles over time, and the need for overzealous sanitization practices. The diverse cultures offer resistance to adventitious or pathogenic cultures and are more likely to remain viable and provide protection through a cheese's entire life span.

The diversity in kefir is its strength! It's also its best attribute for another reason: The diversity of its microorganisms contributes a diversity of flavors in cheeses that is as good as, if not better than, many commercially available raw milk cheeses.

Keeping Fungal Cultures

Two fungal cultures that can be kept easily at home for making white-rinded and blue cheeses are *Geotrichum candidum* and *Penicillium roqueforti*.

Geotrichum, indigenous to raw milk, is a fungal culture that plays an important role in the development of white-rinded and washed-rind cheeses. Raw milk cheesemakers can ensure that their cheeses will develop white *Geotrichum* rinds by creating the conditions that the native milk fungus needs to thrive. A cheesemaker who uses pasteurized milk can rely on kefir culture to provide abundant *Geotrichum* spores to their cheeses.

Maintaining a kefir culture is all you have to do to encourage the *Geotrichum candidum* fungus. Among all the diverse bacterial cultures in kefir are several different fungal species, including *G. candidum*. Kefir, left to ferment undisturbed for one week, will grow a healthy white coat of this beneficial cheese fungus on its surface. And if kefir is used as a starter culture for making a Camembert, the cheese will have the culture it needs to grow a healthy *Geotrichum*-influenced rind. More information on keeping *G. candidum* will be provided in chapters 12 and 16.

Penicillium roqueforti, the fungus responsible for making cheese blue, can also be cultivated easily at home—on a piece of moldy bread. Blue-cheese fungus is the ultimate counter-culture: It's easy to grow on a piece of sourdough bread (see chapter 17), and once you've grown it and dried it out, the dry fungal spores will last for years without any care. Whenever you need a little bit of fungal culture to make a batch of blue cheese, simply break off a small piece of the moldy bread, dissolve it in some cold water, then pour the water with its *P. roqueforti* spores over your ripening milk.

Keeping Washed-Rind Cultures

Brevibacterium linens is an important player in a microbial community that ripens washed-rind cheeses. This culture, to be discussed in greater detail in chapter 18, helps washed-rind cheeses develop their vibrant orange color and funky odors.

Kefir grains floating atop a week-old batch of kefir show the wrinkled growth of *Geotrichum candidum*, the fungus that gives aged chèvre cheeses their distinctly textured rinds.

Brevibacterium linens, a common soil organism that grows in perpetually damp environments, loves to grow on perpetually wet washed-rind cheeses. A ripe washed-rind cheese is therefore an excellent source of the *B. linens* community, and these cultures can be propagated from one cheese to another on the fabric of the cloth that's used to regularly wash the rinds of the older washed-rind cheeses. Every fresh cheese that's washed with that rag will then be inoculated with *B. linens* culture, and continued washing will ensure that the conditions are right for the continued growth of this ripening culture.

CHAPTER FOUR

Rennet

Rennet is used as a catchall term for several different enzymes that help milk separate its curds. Though often labeled one and the same, these milk-coagulating enzymes come from very different sources, some more unsavory than others.

Considering their origins, I believe that their labeling standards don't give cheese eaters or cheesemakers enough information to chew on. I've decided, therefore, to set my own standard in this book.

So when I mention rennet, I am referring to true rennet: the enzymes, primarily chymosin, derived from young animals' stomachs that coagulate milk. More specifically, if it comes from calves, I'll call it calf rennet; if it comes from kids (goat kids, that is), I'll call it kid rennet.

Vegetable rennet is another term I will use, referring to extracts of plants, including cardoons and figs that have been traditionally used by cheesemakers to set their curds.

Microbial enzyme is a third term I will use, referring to the products of certain natural fungal cultures that have a similar coagulating effect to rennet when added to milk. These enzymes are often labeled as vegetarian, though that title may be misleading.

The last type of enzyme that I will reference is FPC—*fermentation produced chymosin*. FPC is also labeled as microbial enzyme, but this label doesn't tell the enzyme's whole story: FPC is produced with transgenic bacteria or fungi containing genetic material of bovine (or camelid) origin that produce

an enzyme functionally and chemically identical to the chymosin in calf (or camel) rennet. So because this enzyme is the same as the one found in calf rennet, but produced with the aid of genetic modification technology, I will call this enzyme genetically modified rennet, or GM rennet.

How Enzymes Work

All of these enzymes work in similar ways: They coagulate milk proteins into curd. Milk proteins take the form of micelles, collections of protein that behave as if they are dissolved in the fluid milk. Milk-coagulating enzymes change that.

Enzymes, like chymosin, fit into their partner proteins much like a key fits into a lock. And when the match is right, the key turns, the lock is opened, and the protein's form changes. Chymosin fits perfectly with casein, the most common milk protein, and when the conditions are favorable (the milk is warm and acidic), rennet changes the shape of the casein, causing it to coagulate, settle out of the milk, and form into curd.

Milk will separate into curds and whey without enzymes if left to sour, but the separation is slow and unpredictable and yields a soft and acidic cheese. Enzymes encourage the natural separation to happen much more quickly, yielding a curd that's stronger, less acidic, and more adaptable to the many different cheesemaking processes. Most cheeses just cannot be made without these coagulating enzymes.

Rennet causes warm and acidic milk to set into a firm curd, seen here as a clean break.

True Rennet

Rennet is actually a mixture of enzymes, including chymosin and pepsin that help coagulate milk into cheese. Of animal origin, rennet is not only useful to cheesemaking but also an integral part of mammalian biology.

Cheese isn't just made in a pot. Newborn calves, kids, and lambs produce enzymes in their stomachs, which transform their mothers' milk into cheese inside their bellies. The fourth stomach of young ruminants, also known as the abomasum, produces enzymes, which, when they come into contact with warm and acidic milk, turn it into a semi-solid curd that is more easily digested. The solid milk can be more easily churned and held by the stomach, and the young animal is therefore able to extract more nutrition from its mother's milk.

It is not just calves, kids, and lambs that produce enzymes to help them digest their mothers' milk, but human infants, too. Proof of the cheesemaking that happens inside the bellies of our young can be seen when an infant spits up on your shoulder: That gelatinous white substance is raw breast-milk cheese that coagulated in the little one's stomach!

Milk and rennet form an extraordinary partnership that is vital to the health and well-being of most young mammals. The two were created for each other, and each would not exist without the other. They represent a sacred relationship between mother and child . . . a bond that isn't necessarily broken by the use of rennet in cheesemaking.

To the dismay of many cheese eaters, using traditional rennet in cheesemaking demands the slaughter of young, milk-fed calves. But those calves were not slaughtered for their rennet, and using calf rennet, as I shall soon explain, may be a more ethical choice than many of its alternatives.

The Discovery of Rennet

As soon as humans starting keeping domesticated animals for milk, possibly as long as 10,000 years ago, simple methods of making cheese without rennet likely evolved as ways to preserve their milk. Recent archaeological findings show the earliest evidence of cheesemaking dating from 7,500 years ago in what is now Poland, but when and where we came across the idea of using rennet to make cheese is as yet unknown.

It may be that rennet cheesemaking evolved independently in each region where goats, sheep, cows, zebu, buffalo, yaks, camels, donkeys, and horses (each of which was kept historically in part for milk) were domesticated. And it's possible that rennet was discovered entirely by chance. The official story of its discovery goes a little something like this:

Somewhere in Asia, many thousands of years ago, a shepherd went out to pasture with her sheep, and as she always did, she took along some fresh sheep's milk in a sheepskin bag. But on this particularly fateful day in history, the shepherd poured her milk into a small bag that she had fashioned from the fourth stomach of a young lamb she had recently slaughtered.

Over the course of the day, the raw sheep's milk warmed in the hot sun. As the milk warmed, the raw milk bacteria slowly began souring the milk; the warm, sour milk, in contact with the rennet enzyme on the lining of the young animal's stomach, transformed into curds; and as the shepherd continued walking, the repeated shaking of the bag broke the curd up into smaller bits, firming them up into cheese.

When the shepherd went to take a drink from the bag later in the day, what she found surprised her: The milk had transformed into strangely delicious and chewy chunks and had left behind a tasty, tangy liquid. She liked what she tasted,

Abomasa, the fourth stomachs of young ruminants, where cheese happens naturally.

realized that it was the young lamb's stomach that had made the milk curdle, and took her discovery back to her village, where she started a cheese-making tradition that continues to this day.

But that's just the official story; there is an unofficial story about the discovery of rennet cheese, one that goes back much further than the domestication of dairy animals. It may very well be that humans have been enjoying cheese for as long as we've been humans, for any hunter that ever killed a young deer or antelope or elephant in the hunt upon cutting open that young animal's stomach would have found fresh cheese curds! These sweet and warm curds were likely the prize of the hunt, and possibly the most delicious and nourishing food to come from an animal.

Microbial Enzymes

Certain fungal cultures, such as *Rhizomucor miehei*, naturally produce enzymes that coagulate milk much like rennet. Used by cheesemakers to make "vegetarian" cheeses, microbial enzymes are also used for kosher- and halal-certified cheeses; they avoid the need to use kosher- or halal-certified calf rennet, a near impossibility to source.

These microbial enzymes, unfortunately, do not make cheese that's as good as what calf rennet creates. These synthetic enzymes can give cheese a rubbery texture and, when aged, an upsetting bitterness. As well, the curd formed by microbial enzymes behaves slightly differently from rennet curd and can create complications, especially if a cheesemaker is accustomed to using calf rennet.

Microbial enzymes also carry ethical baggage, and using them in cheesemaking may not necessarily make a more virtuous cheese. Folks who wish to take more control of their food by making their own cheese may be unknowingly relinquishing control to corporate interests if they use these coagulants, for many of the companies that manufacture microbial enzymes are wholly owned subsidiaries of agribusiness and chemical corporations. (Marzyme—trade name Marschall—the most common microbial enzyme tablet, is manufactured by Danisco, a subsidiary of DuPont.) As well, a cheese made with a "vegetarian" enzyme can be no more vegetarian than milk, which is hardly a vegetarian food.

"Vegetarian" Enzymes

Until the 1970s the majority of cheeses produced in North America were made with calf rennet. Derived from calf stomachs, calf rennet is inextricably linked with veal production, but a consumer revolt in the 1960s and '70s against the practice of keeping veal calves in pens (a practice that's linked to keeping dairy cattle confined—another consequence of industrial dairying) gutted the veal industry.

A decline in the availability of vells (the term used for salted and cured stomachs) drove the price of calf rennet up. Spurred to search for alternatives to the now costly coagulant, food scientists developed microbial enzymes, produced by isolated fungal cultures, which could essentially replace rennet in cheesemaking. Microbial enzymes soon became the standard coagulant used in North American cheese production.

Cheeses made with these natural microbial enzymes are often described as "vegetarian," but this label fails to contextualize North American dairy practices. The unfortunate truth of cheese is that there is no such thing as a vegetarian cheese, because milk production is dependent on animal slaughter. And using vegetarian rennets in cheesemaking does not avoid the slaughter of animals, for calves are not slaughtered for their rennet, nor are they slaughtered for their veal: They are slaughtered for their mothers' milk.

To produce milk, dairy cows must give birth to a calf (or two) every year. Half of the calves born—the males—do not have a role to play in the dairy. But most of the female calves are not needed, either: A typical dairy cow will give birth to many calves over her productive life, but only one calf is needed to replace her at the end of her days. Because the remaining calves are bred for milk production, they are often considered not worth raising to adulthood for meat and therefore are not of value to the dairy.

It used to be that these young "bobby" calves were fed their mothers' milk, and raised to a certain size, then slaughtered for veal. But as consumers' distaste for veal grew, farmers found themselves unable to sell their calves. However, the dynamics of dairying didn't change, and dairy farmers still don't need most of the young calves born into their dairies. Dairy farmers now often find themselves with few other options but to slaughter their bobby calves at birth.

So a protest against inhumane calf-raising practices has resulted in the needless slaughter of newborn calves at dairy farms across North America. Harvesting rennet from milk-fed calves doesn't sound so bad, in context, but unfortunately, because very few calves are now raised for veal in North America, most calf rennet available here is sourced from New Zealand and Australia, two dairying countries that have, for now, maintained their veal industries. So-called local artisanal cheeses made with traditional calf rennet are therefore made with an integral enzyme that was manufactured and shipped from halfway around the world!

Calf rennet is the most traditional and natural way of transforming milk into cheese and thus aiding its preservation. The use of calf's rennet can also be a part of a "full-circle" approach to cheesemaking in a way that microbial rennets cannot. So if you make cheese, or eat cheese, or just drink milk, maybe it's time to reconsider veal: Ethically raised, milk-fed veal is becoming more and more available to consumers, and supporting farmers who raise ethical veal can help promote sustainable dairying and ensure a healthy supply of locally produced calf rennet.

I believe the time is right for North American cheesemakers to take back their rennet . . . especially considering the increased prevalence of genetically modified rennet.

Genetically Modified Rennet

Made with the aid of genetically modified bacteria and fungi that produce enzymes in bio-reactors, genetically modified rennet is not limited by the dwindling supply of calf stomachs that has resulted from the decline of the veal industry.

FPC is produced with the aid of transgenic bacteria, specifically species of *Escherichia coli* (*E. coli*) or *Aspergillus niger* that have been given a gene from a cow expressing the production of chymosin, which allows the microorganisms to synthesize the enzyme in bio-reactors without the slaughter of calves.

This bioengineered rennet is now used in the manufacture of the vast majority of North American cheeses. The vast majority of North Americans, however, do not know that their cheeses are being made with a product of genetic engineering technology, as cheeses made with these rennets list only "microbial enzyme" on their packaging.

FPC manufacturers do not consider their enzymes to be genetically modified, and cheesemakers who use these enzymes generally agree. Because, they say, the enzyme itself is produced with unmodified cattle genes, the enzyme is precisely the same as chymosin sourced from calf stomachs and works in exactly the same way. Though the enzymes are produced with the aid of bacteria that are genetically modified, the enzymes they produce are not explicitly genetically modified (they are the same chymosin found in calf stomachs) and therefore do not need to be labeled as genetically modified. And the enzymes themselves are not GMOs because they do not contain any residuals of the genetically modified organisms.

Chr. Hansen, a leading manufacturer of FPC, states this perspective explicitly on their website:

> Many food and medical enzymes and additives such as citric acid, aspartame, fructose, lipases and Fermentation-Produced Chymosins are produced using GMO technology . . . Despite the fact these enzymes and additives are produced with the assistance of GM technology, the products are not genetically modified and are separated from the host organism. As a result, all Chr. Hansen enzymes are considered GMO-free.

Considered by whom to be GMO-free? By food manufacturers, by the biotechnology industry, by many governments and their GMO labeling laws, perhaps, but not by the discerning public, nor by organic certifiers, who do not allow any products of GM technology to be used in organic agriculture or organically produced foods (although there are exceptions to this rule).

The concern about genetic modification is much greater than the question of the effect of consuming GMOs: It is about the normalization of a technology that is having enormous economic, social, and environmental consequences; and about the inclusion of the products of such technology in our food without our consent. GM technologies are not being given adequate testing prior to release (GM rennet was not subject to a review when it was given Generally Regarded As Safe [GRAS] status by the US Food and Drug Administration), and though most consumers want to know if GM products are in their food, legislation to require their labeling has been largely stalled thus far by powerful and well-financed corporate interests.

So how is a cheesemaker or consumer to know if the microbial enzyme in their cheese is GM rennet if labeling laws don't require such information to be disclosed? GM rennets are often found labeled as 100 percent pure chymosin, or with the brand name CHY-MAX. However, if you're concerned about the many consequences of biotechnology, I'd suggest taking a cautious approach to "microbial rennets" in general: Microbial enzymes are either produced with genetic engineering technology or manufactured by corporations that promote genetic modification—DuPont, producer of one popular brand of non-GM microbial rennet, is also the world's leading developer of genetically modified seeds.

Vegetable Rennets

Vegetable rennets are coagulating agents of plant origin. In places where cheesemakers could not afford the high cost of rennet, they traditionally used certain plants with milk-coagulating abilities to make their cheeses.

As an alternative to the use of animal rennet, microbial enzymes, or GM rennet in cheesemaking, you can grow your own coagulants by cultivating these plants. However, if you choose to use these rennet alternatives, keep in mind that they are not as effective as other milk-clotting enzymes and do not allow a cheesemaker to handle the curds in exactly the same manner. Unfortunately, commercial suppliers of these vegetable rennets are difficult to find; if you wish to use them, you'll have to prepare them yourself.

Cardoon

The most famous of milk-clotting plants is the cardoon, whose botanical Latin name is *Cynara cardunculus*. A close relative of the globe artichoke, cardoon is grown commercially for its succulent stems, which are popular in Mediterranean cookery, and also as an ornamental for its striking flowers that bristle with thorns and are crowned with a cluster of bold blue petals. It is the dried petals of cardoon that yield a milk-coagulating enzyme used to make some traditional Portuguese and Spanish sheep and goat cheeses such as Torta del Casar. A decoction of dried petals can be made that, when poured into milk, will form a semi-firm curd that can be handled much like rennet curd. To do so, let 2 grams of dried cardoon petals steep in ¼ cup (60 mL) of warm water for an hour. Strain off the liquid, squeezing the petals of their juice, and pour the rennet substitute over a gallon of warm acidic milk exactly as you would calf rennet.

Using Rennet

Rennet is added to the milk toward the beginning of the cheesemaking process. In the making of most cheeses, the enzyme is added in the following manner: First, the milk is warmed to 90°F (32°C); next, the starter cultures are added to the milk; the cultured milk is incubated for an hour so that it begins to sour; then the rennet is added to the warm, slightly sour milk.

Rennet is a very reactive ingredient, and only a small amount is needed to set the curd. The rennet should be measured out to a fairly exact quantity, according to the recommended dosage and the amount of milk. For example, the makers of the rennet tablets I use recommend one tablet to coagulate every 5 gallons (20 L) of milk. According to that dose, about one-quarter of a tablet is needed to set 1 gallon of milk.

But because different cheeses need different rennet doses, the amount of rennet added to the milk can vary widely. One of the most important differences between soft and hard cheeses is the amount of rennet used: Semi-firm cheeses such as blue cheese and mozzarella have a regular dose of rennet; firm cheeses such as Alpine cheeses have a double dose of rennet; and soft cheeses such as chèvre use about a quarter of the recommended amount. So when using the same brand of rennet as above, only one-sixteenth of a tablet would be needed to make a 1-gallon batch of chèvre, whereas two tablets would be needed to make a 5-gallon batch of Alpine cheese.

Because rennet is so potent, it must be diluted in water before being added to the milk. The water used to dilute the rennet should be clean—water can be a source of coliform bacteria that can cause the curd to float and the cheese to bloat—but also not chlorinated; chlorine, an oxidizer and antibacterial agent, can affect the activity of the coagulating enzymes as well as the microbial development of a cheese. Most types of chlorine can be evaporated from water by leaving the water to stand, uncovered, at room temperature for one day.

Once the rennet has been added to the milk, the renneted milk is usually left to set for an hour. However, depending on the dose of rennet, but also on the acidity of the milk (fresher milk will take longer, while older milk will set more quickly), the curdling can take place in as little as 15 minutes and as much as two days. Either way, the milk should be left undisturbed as it sets to ensure the development of strong curds.

The milk should also be kept warm as the curd develops. Rennet sets best and most quickly at around body temperature (its preferred conditions are, after all, inside an animal's stomach); so incubating the milk to keep it around 90°F (32°C) helps the curds form in a timely manner.

Rennet Types

Rennet is available from calves, kids, or lambs. However, each of these animals' rennet is not isolated to the same species, and all of them will universally coagulate any type of milk. Some cheesemaking designations require that goats' milk cheeses be set with kid rennet, but no significant differences would result in the cheese if calf rennet were used in its place.

Rennet comes in liquids, tablets, powders, and pastes. Each form of rennet works in a similar manner, and each one can be used in place of the other, but each has its own specifications and dosages: be sure to consult the instructions on the package before using.

Liquid rennet is the most convenient of all rennets to use, as it is easy to measure, dissolves quickly in water, and can be added to milk in a flash. However, liquid rennets often contain preservatives to control microbial growth. Usually not labeled along with the "enzyme" in cheese, liquid rennets usually contain unexpected ingredients, such as sodium benzoate.

Powdered rennet is made of purified, dried chymosin and pepsin but also contains salt, a result of the particular processing that makes it. Powdered rennets offer the purest natural form of chymosin, do not generally contain additives, and have the longest shelf life.

Rennet tablets are made of pressed rennet powder. They have the same qualities as powdered rennet, and are used in the same way, but are made to be easy to measure for small batches of cheese and are most suited to home cheesemaking. Rennet tablets are what I use in my cheesemaking, and the recipes of this book include instructions for a particular brand of tablets known as WalcoRen, now available from most online cheesemaking supply stores.

The rennet I recommend for small-scale natural cheesemaking.

Rennet paste is a relatively unprocessed form of rennet. It is essentially made of salted and cured calf stomachs (also known as vells), rehydrated in water. It contains not only chymosin but also high levels of pepsin and lipase (along with preservatives to keep it fresh). Cheeses made with rennet paste tend to develop more interesting flavors due, in part, to its lipase. An enzyme found in raw milk and unprocessed rennet, and produced by many species of bacteria, lipase aids in the breakdown of fat and creates fatty acids that improve the taste of many traditionally made cheeses. Many pasteurized milk cheeses have lipase added to the milk (sourced from cows' pancreases) to give them complex flavors that they would otherwise lack.

You can even use the vells directly, much as cheesemakers have for thousands of years (instructions for their use are included at the end of the recipe for rennet). Once upon a time, purchasing calf vells for rennet was commonplace (folks would buy them at their local tripe shop!); these days, however, vells are not generally available for sale to the public in North America. If you wish to go back to the source, you'll have to harvest and process the vell yourself.

- RECIPE -
RENNET

The preparation of vells for cheesemaking is intimately related to the making of biodynamic preparations. Much like cows' horns filled with manure and left to cure in the earth from solstice to solstice (Biodynamic Preparation 500), or yarrow blossoms stuffed into stag bladders and aged from fall to spring (BP 502), the young calf stomachs are prepared in a careful manner, cured for six months, and even stirred into milk slowly and intentionally in the same way that biodynamic preparations are added to water. And in much the same way that biodynamic preparations harness life forces, encouraging them to catalyze the health and wholeness of

a farm, the cured calf stomachs are used as a catalyst to transform milk into cheese as part of a whole system in an almost spiritual invocation.

To make your own rennet, you will need to slaughter a very young calf, kid, or lamb. If you aren't raising your own animals, get in touch with local dairy farmers and find out if they have bull calves or bucklings available for sale. Cattle dairies will often have calves throughout the year, but kids and lambs are generally born only in the spring.

Raise the calf on its mother's milk, preferably by her side. If you don't have its mother's milk to raise it, be sure that you feed it good milk. A young calf, much like an infant, will have trouble digesting pasteurized, homogenized milk from the grocery store because it won't form strong curd when it comes in contact with the rennet in its stomach.

When the animal is slaughtered, it should not yet be eating grass. The presence of grass in the digestive tract can make the processing of the abomasum more challenging. As well, the enzyme contents of animals' stomachs change as they grow: The older the animal, the less chymosin will be in its stomach, and the less coagulating power the vell will have.

Keep the young animal off milk overnight before slaughter; otherwise, there will be curds in the fourth stomach. Traditional Sardinian cheesemakers make a cheese from this curdled milk that's aged in the stomach, but for rennet processing, these curds just get in the way!

As a word of warning before you proceed with making your own rennet: Selling cheese made with rennet prepared in-house is generally not permitted. Even though the use of home-butchered organs is permitted for making certain biodynamic preparations that are then applied to the soil; even though using vells prepared in-house would enhance the unity of a cheesemaker's operation; and even though milk and rennet form an important partnership, and are almost one and the same, cheesemakers are not permitted to use their own calves' stomachs for the production of their cheese. Only certified commercially prepared rennet may be used in the manufacture of cheeses to be sold to the public.

Rennet

Ingredients

1 young, unweaned calf, kid, or lamb
Good salt (see chapter 5: Salt)

Equipment

Well-sharpened slaughtering, butchering, and
  skinning knives
Stainless bowls

Cutting boards
2 small lengths of string
Hand pump

Time Frame

1–2 hours to slaughter and butcher the animal and
  prepare the vell for curing; 6 months to cure
  the vell

The abomasum, located just upstream of the intestines, is the true stomach and the original source of rennet.

YIELD

1 vell

TECHNIQUE

Slaughter the animal: According to your religious
authority, or according to practices that you see
as humane and ethical, take the life of the young
animal. It's a good idea to have an experienced
slaughterer on hand to assure a pain-free
passing and an experienced butcher to carefully
carve the carcass.

Butcher the animal: Post-slaughter, hang the
animal, skin it, then carefully slice open its
abdominal cavity to avoid puncturing any
organs within. Cautiously cut off the gall-
bladder; puncturing and spilling its contents
can taint the meat as well as the rennet.

Find the abomasum: Carefully remove the
digestive tract, and place on a table to examine.
Follow the progression of the tract until you find
the fourth stomach also known as the true
stomach, or abomasum, which will be just
before the intestines.

Cut out the abomasum, including the sphincter
between it and the previous stomach, and 5
inches (12.5 cm) of the intestines downstream
of the bottom sphincter. Do not clean out the
stomach contents—the rennet is not just in the
lining of the stomach but also in its contents. Do
not cut or puncture the stomach itself, as it will
need to be inflated later in its processing. If you
are using the stomach of an older animal, it will
be filled with digesting grasses and will have to
be cleaned with water—and the coagulating
power of the vell may thus be lessened.

Salt the abomasum: Cover the stomach well with
fine, additive-free salt.

Cure the abomasum: Leave the abomasum to
cure, covered in salt, in a closed container for at
least 6 months. This long curing period activates
the enzymes in the stomach. Once cured, the
vell can then be dried.

Inflate the abomasum: Tie off the sphincter on
the top end of the stomach with a piece of

To dry the cured abomasum, inflate with a pump, and
leave to air-dry.

string, inflate the stomach from the intestine (using a pump), then tie off the bottom sphincter to keep the air in.

Dry the abomasum: Hang the inflated stomach to dry for a week in a drafty and dry space. Once dried, the vell can be kept in a sealed container and will remain active for years.

Use the vell: About ½ inch square (about 1 square cm) of vell is usually all that is needed to set 1 gallon of milk (4 L), though the amount to use depends on the concentration of enzymes in the stomach, which is dependent on the age of the animal at slaughter.

To use the vell, cut off a small piece of the stomach and add it to a small amount (less than ¼ cup [60 mL]) of lightly salted active whey or water. Leave the vell overnight to steep overnight, and in the morning remove it from the water and grind it into a paste. You then mix it back into the water or whey and add it to the milk much like other rennets.

– RECIPE –
JUNKET PUDDING

To demonstrate the effects of rennet on milk, here's a recipe for a dessert known as junket: a Jell-O-like pudding made by setting milk, often sweetened and spiced, with rennet. Junket was once served in many North American and British homes, but today it is just a hazy childhood memory of Baby Boomers.

Junket is a victim of the pasteurization and homogenization of our milk supply. The simple recipe for junket does not work with overprocessed milk; pasteurized milk does not contain the bacterial cultures that naturally sour the milk and allow the junket to set. And junket made with homogenized milk just does not set a strong curd.

Historically, junket would be made at home with raw milk, delivered by the milkman daily; a small

piece of a dried calf's stomach would be picked up from the tripe shop and used to set the curd. After World War II, raw milk deliveries stopped, and tripe shops closed up. Homogenization of our milk supply dealt the final blow.

I think it's about time for junket to make a comeback, for it's a delicious dessert, with a fantastic texture. Junket can be flavored with a variety of milk spices, such as cinnamon, nutmeg, and cardamom, or allspice, ginger, and mace. It can be made with any type of milk, be it cow, goat, or sheep. And it can be spiked with cream or spirits to make a more celebratory dessert. Junket is also much more easily made than other milk custards, and does not require careful cooking and stirring to avoid curdling.

If you want a natural, healthful food to transition your toddler off breast milk, junket may be just what you're looking for. Not recommended for feeding infants by contemporary nutritionists (due to systemic fear of raw milk and homemade dairy products), junket (essentially partially digested milk) was commonly fed to children, invalids, and people with digestive tract ailments. It was even served in hospitals that often kept a well-tended herd of cattle just to make it, to encourage good nutrition and healing. (How that evolved into the blue Jell-O served at hospitals today is beyond my understanding.)

If you are making junket with raw milk, follow the recipe as written. If your only option is pasteurized milk, you'll need to include an extra culturing step, as pasteurized milk does not contain populations of good bacteria that help acidify the milk, which is necessary for ensuring a good rennet set, as well as for keeping the growth of unwanted microorganisms in check. If you only have pasteurized, homogenized milk, I suggest seeking out better milk.

INGREDIENTS

1 quart (1 L) good milk, raw or pasteurized but not homogenized
1 tablespoon (15 mL) active kefir or whey (optional if using raw milk)

To make junket, add sugar and spice to warm milk; mix in rennet; then ladle the sweetened, spiced, and renneted milk into cups to set.

Sugar or any other sweetener to taste
   (¼–1 cup [60–240 mL], or none at all)
¼ teaspoon (1 mL) salt
1 teaspoon (5 mL) ground cinnamon
½ teaspoon (2 mL) ground nutmeg
½ teaspoon (2 mL) ground cardamom
Regular dose rennet (I use ¹⁄₁₆ tablet WalcoRen
   rennet for 1 quart milk)

EQUIPMENT

Small pot
4 serving cups or bowls

TIME FRAME

1 hour

YIELD

Makes 1 quart (1 L) junket pudding

TECHNIQUE

Warm your milk to baby-bottle-warm,
   about 90°F (32°C).
Culture your milk (optional for raw milk) by
   adding the kefir or whey to the warm milk.
   Leave the pot to incubate for 1 hour, to help the
   bacterial cultures flourish.
Mix the sugar, salt, and spices into the milk well.
Add the rennet: Dissolve the rennet in ¼ cup (60
   mL) of water, then add it to the milk. Mix the
   rennet in with a gentle stirring.
Pour the renneted milk into individual cups
   immediately after adding the rennet. This way,
   the curd will form a firm set in the cups.
Let the pudding set, covered, at room temperature for 1 to 2 hours until firm. Junket is best as
   soon as it has set but can also be kept, refrigerated, for several days.

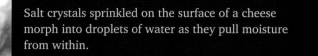

Salt crystals sprinkled on the surface of a cheese morph into droplets of water as they pull moisture from within.

CHAPTER FIVE

Salt

Salt is an essential tool that cheesemakers use to make their cheeses. Integral to the preservation of many traditional foods, salt plays an important role in preserving cheeses as well. Salting has numerous beneficial effects. By controlling the moisture content in cheese, salting firms up its flesh, forms protective rinds, controls unwanted fungal and bacterial growth, and slows and mellows cheeses' aging.

Cheese just isn't the same without salt. Not a single cheese can be aged without it; even fresh cheeses don't last as long or taste the same if they haven't been salted. Cheeses require a surprising amount of salt. Folks who have trouble adding salt to their food may be a bit squeamish about the amount needed to properly salt a cheese. But cheeses truly need as much salt as they do, and skimping on it can have ugly consequences!

How Salt Works

Salt gets the moisture out of cheese.

A visualization of salt's powers can be gracefully demonstrated by sprinkling fine salt on the surface of a ripe Camembert. Try it yourself: Take a pinch of fine-crystalled salt, sprinkle it over the top of the cheese, and leave the salted cheese out at room temperature. After 10 minutes you will observe the transformation of the salt into small droplets of water.

As the salt crystals sit in contact with the rind of the cheese, they begin to draw moisture toward

them. The crystals of salt begin to transform into small beads of moisture, actually whey, pulled from within the cheese.

What you are witnessing is osmosis, a natural interplay of salt and water that influences this exchange between the cheese and the salt. The salt crystals, with less moisture than the cheese, pull the cheese's moisture toward them. And as the salt dissolves in the moisture pulled from the cheese, some of the salt is absorbed into the Camembert.

The movement of salt and moisture between the salt and the cheese continues until they come to a balance where the cheese is as salty as the drops of water. This balance is known as an osmotic equilibrium, and once it is achieved, the salt has done its work: The cheese has less moisture and more salt, two conditions that help cheeses age.

The Effects of Salting

Salting helps make a firmer curd. Cheese isn't exactly a solid; it is classified as a semi-liquid gel. Cheese flows and changes its shape as it ages; salting pulls moisture out of a cheese and slows its flow.

Salting forms rinds on the surface of cheeses that protect them as they age. A Camembert without a strong rind will spill its softening curd into a puddle as its interior liquefies with age; rindless fetas and mozzarellas will melt into their brines.

Salting preserves cheese by restricting the growth of unwanted bacteria and fungi. Most species of bacteria and fungi are not tolerant of high levels of salt or low levels of moisture and are kept in check by a proper salting. Interestingly, all of the beneficial bacteria and fungi that help cheeses age are tolerant of a certain amount of salt; indeed a proper salting encourages their growth over other, less beneficial microorganisms.

Salting also slows the aging of cheese. In mildly salty conditions ripening cultures grow more slowly and cause cheeses to mature in a mellower manner. Salting helps cheeses age over long periods of time and develop their characteristic flavors and textures.

Undersalting Cheeses

My first attempt to make blue cheese was plagued by a problem that resulted from insufficient salting. Not quite understanding the important role salting plays in preparing cheeses for their aging journey, I salted my cheese like I did my dinner, sprinkling just a pinch atop my draining curd. I then let the cheese dry for a day and put it away to age in my cave.

I first noticed something was wrong when a fine fuzz started appearing on my cheese. Within a few days a strange fungus had completely covered its surface and was even growing on its draining mat. My first blue didn't turn out blue at all, but rather blackish gray.

Consulting a French cheesemaking guidebook that took a liberal view toward fungus, I learned a few things about what was growing on my cheese. The name of the fungus was cat fur fungus (also known as *Mucor*), and what a suitable name that is, for the blackish-gray "fur" grew in tufts to over an inch in length.

I read on. According to the guide, cat fur grows on cheeses that are insufficiently salted, and thus too moist; proper salting will eliminate its growth. The guidebook had some more advice for me: The cat fur fungus is nontoxic and does not affect the flavor of the cheeses it grows on—it is merely an aesthetic problem. The book advised me to pat the cat's fur down and eat the cheese!

Following the guidebook's unlikely guidance, I patted down the fur, and, apprehensively, I tasted my "blue" cheese. And I'm sure glad I did, because it was fantastic! Had I just thrown it away, I would have never known that my first blue cheese wasn't a complete failure. I now know that proper salting can eliminate the growth of cat fur on cheese, and I've never found the fungus growing again on any of my cheeses.

Oversalting

Of course, you can also oversalt your cheeses. Oversalting can change the ecological balance of a cheese, discouraging certain species while encouraging others. *Penicillium roqueforti*, more tolerant of salt than other species such as *Geotrichum candidum*, can cause oversalted white-rinded cheeses to ripen into blues.

A cheese that is extremely oversalted will have too much moisture drained from it, and the dry, salty conditions within the cheese can impede all bacterial and fungal growth. Such a cheese will not support the microorganisms necessary for its development and therefore will not age.

There is, fortunately, a certain range within which salting works well. Cheeses don't have to be precisely salted to the milligram; a haphazard approach—using pinches, dashes, and rough tablespoons—is my preferred way to salt, and my cheeses don't suffer. Of course, if you aren't an old hand at salting cheeses, it is a good idea to carefully measure out your salt according to the recipe's recommendations.

Good Salt

Any salt that is fine-crystalled and additive-free will work for cheesemaking. Good salts to use include sea salt, pickling salt, kosher salt, and cheese salt.

Household table salt is not suitable for cheesemaking as it contains iodine; iodine, used as an antibacterial agent to sterilize wounds, can also sterilize your cheeses. As well, table salt contains anti-caking agents that absorb moisture and help it flow smoothly and are not a suitable ingredient in cheese. Coarse-crystalled salt is also not a good choice for cheesemaking as the large and heavy salt grains will not stick to the cheese.

Gray cat fur and other unwanted fungi growing on a
poorly salted cheese.

Sea Salt

My favorite for flavor, sea salt is an excellent salt for cheesemaking. Not just sodium chloride, sea salt contains other minerals present in the seawater from which it is evaporated and thus contributes additional nuances and nutrition to cheese.

Unrefined sea salt can leave a slight staining on the surfaces of cheeses. This coloration is no reason to avoid its use; it is a sign of the high mineral content in sea salt, which contributes to the healthful benefits and fine flavor of sea-salted cheeses.

Unrefined sea salt is also home to a community of salt-tolerant bacteria. When sea salt is added to cheese it is likely that some of these salt-tolerant bacteria will establish themselves upon their rinds and help the cheeses develop more complex flavors.

Mineral Salt

Refined or unrefined mineral salts are another good choice for salting cheeses.

These salts are mined, rather than evaporated from seawater, and are the most common salts available in grocery stores. Often sold as rock salts or pickling salts, they are relatively inexpensive and work just fine for cheesemaking, so long as they are finely ground and additive-free.

Unrefined, mineral-rich rock salts can leave behind colorful marks on the rinds of the cheese: Pink Himalayan rock salt, for example, will leave a pinkish blush on a fresh cheese.

Kosher Salt

Kosher salt is a mineral salt with a complex crystalline structure. It is prepared by subjecting dissolved rock salt to a careful evaporation that results in salt grains with a fine, pyramid-like form.

Kosher salt's traditional use is in the Jewish practice of koshering meat, wherein salt is applied to the surfaces of a carcass to help pull the blood out of the flesh. Kosher salt's unique crystal structure helps it adhere to the meat; this crystal structure also makes it particularly well suited to adhering to a cheese and pulling out its whey.

Kosher salt's light structure means that a greater volume of this salt must be added to cheeses to have the same effect as other salts. Be sure to increase the volume of salt added to cheeses by 50 percent (but don't adjust the weight!) if you use kosher salt.

Cheese Salt

Cheese salt is a highly refined and very fine-crystalled rock salt that is used by many cheesemakers. It is

Three good salts for cheesemaking (left to right): unrefined sea salt, flaked kosher salt, fine cheese salt.

un-iodized and has no added anti-caking agents. This is a utilitarian salt that works well for cheesemaking and can be purchased from cheesemaking supply shops.

Flavored Salts

Flavored salts can be used to impart interesting flavors to your cheeses. For example, smoked salts can give smoky nuances to a cheese, and bacon-flavored salts can make your cheeses taste like bacon. Use these salts to your discretion, and preferably on fresh cheeses.

The Salting Process

There are several different methods of salting cheese. Cheeses can have salt applied to their surfaces, the salt can be mixed in with the curd, or cheeses can be placed into a saturated salt brine. Regardless of which method is used, salting is always carried out toward the end of the cheesemaking process.

When to Salt

Most of the steps involved in making cheese are aimed at getting the moisture out of milk in an attempt to preserve it. From adding rennet to cutting the curd, stirring the curd, and straining the cheese into a form, the goal of a cheesemaker's work can be seen as getting the whey out of cheese. Salt is added to a cheese only after cheesemakers have done everything they can to get as much moisture as possible out of the curd; only then does salt effectively pull the remaining moisture out.

Salting therefore happens toward the end of a cheese's make, usually once a cheese has been formed—though there are some exceptions. Salting is the final step in getting moisture out of a cheese after it has been strained from the pot, formed, and pressed. After salting, a fresh cheese is ready to eat, and a cheese to be aged is ready to be put in the cheese cave to ripen, but only once it's been drained and air-dried.

Draining and Air-Drying

Salting is only effective if the salt is allowed to pull the moisture from the cheese. After a cheese is salted, the cheese must be handled in the right way to allow the salt to do its good work of extracting the whey. After salting, hung cheeses such as chèvre and yogurt cheese are left to hang to allow the salt to remove and drain off more whey; formed cheeses such as Camemberts are placed on a draining rack to drain and dry for a day or two after salting. Flip them as they drain to ensure that they dry evenly on all sides.

Cheeses are generally ready for what's next (whether it be eating or aging) only once they are fully drained and dried and are no longer shiny; if a cheese still seems wet, it should be left to drain and dry a few hours longer; additional salt can be added if the cheese seems especially wet.

Surface Salting

The simplest way to salt a round of cheese is by applying salt to its surface in a process known as . . . surface salting. To surface-salt a cheese, a certain amount of salt is measured out based on the weight of a cheese, and that salt is rubbed into all sides of the cheese.

Salt is added at about 2 percent by weight. So 2 percent of the weight of a 1-pound cheese would be about ⅓ ounce of salt, which measures out to about 1 tablespoon (that's 10 g or 15 mL of salt for 500 g of cheese). Accordingly, a 5-pound cheese will need about 5 tablespoons of salt (a 2.5 kg cheese will need 75 mL of salt).

I generally estimate the amount of salt to add to a cheese from the amount of milk that made the cheese. One gallon of milk makes about 1 pound of cheese; since 1 pound of cheese needs 1 tablespoon of salt, the cheese from 1 gallon's (4 L) milk needs 1 tablespoon (15 mL) of salt. So if I make three Camemberts from a gallon of milk, each one will need ⅓ tablespoon (5 mL) of salt. And if I make a Gouda with 5 gallons (20 L) of milk, 5 tablespoons (75 mL) of salt will effectively salt it.

Be sure to salt all sides of a cheese evenly. If a Camembert is only salted on its top and bottom, the sides may remain too moist and could play host to unwanted fungal growth.

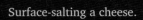
Surface-salting a cheese.

When brining, be sure to cover all exposed parts of a cheese with salt.

Larger cheeses, as well, can be difficult to surface-salt, as they have a smaller surface area for their large size, and their surfaces can only hold a certain amount of salt. A very large wheel of Alpine cheese, for example, may have to be surface-salted two or three times to be effectively salted. Larger cheeses are generally brined at commercial cheese-making operations for this reason.

Brining

Cheeses can be more easily, efficiently, and accurately salted by leaving them in a super-saturated salt brine for a certain amount of time.

It may seem counterintuitive to put a cheese into a liquid to dry it out—but the liquid brine is so salty that osmosis causes it to pull moisture toward it from the cheese. In the process the cheese absorbs some salt from the brine.

Salt brines are prepared from fresh, sweet whey, straight from the cheesemaking pot. A brine is best prepared while the whey is still warm; the salt dissolves much more quickly, and the whey doesn't have time to develop acidity that might sour a cheese.

To make a saturated salt brine, dissolve 2.25 pounds (1 kg) or 3½ cups (840 mL) of salt in every gallon (4 L) of whey, then add 1 or 2 extra cups (240–480 mL) of salt to serve as a salt reservoir. Stir the whey for a minute or two until much of the salt is dissolved. Then place the brine to cool in a refrigerator or a cheese cave.

Leave soft cheeses in the cool salt brine for 2 hours for every pound of cheese (that's 4 hours for every kilogram). A Camembert weighing ½ pound (225 g) will spend 1 hour in the brine, whereas a Brie weighing 2 pounds (1 kg) will spend 4 hours in the brine. Be sure to leave yourself a reminder for when the cheeses are ready to be removed.

Harder cheeses spend more time per pound in their brines. Goudas and cheddars and other firm cheeses spend 3 to 4 hours per pound depending on their firmness. A 40-pound (20-kg) wheel of Gouda would need to spend 120 hours in the brine, whereas a 100-pound (50-kg) wheel of Parmigiano Reggiano needs over 400 hours of brining—that's almost 20 days.

If any parts of a large cheese are exposed to air when the cheese is in the brine, be sure to cover those parts in salt. If a large cheese is left in brine for multiple days, flip it in its brine every day to be sure that all sides are sufficiently salted.

Brines need a little bit of maintenance to be kept in good shape. They should be kept in a closed container in a cool place, either a refrigerator or a cheese cave. And as the salt reservoir at the bottom of the brine is depleted (every time a cheese is placed into the brine, some of the brine's salt is taken up), it will need to be topped up. You can use a brine for many months so long as it's well maintained. And as long as there is some undissolved salt in the brine, the brine will work effectively.

Cheesemakers generally keep separate brines for separate cheeses to prevent cultural mingling. A brine that's used for blue cheese or made from blue cheese whey, for instance, will have *Penicillium roqueforti* fungal spores that could infect a Camembert and give it blue freckles. I find, however, that such concerns about contamination are unnecessary if you work with strong natural cultures and are careful about creating the conditions each cultural regime needs to thrive.

Salting the Curds

The softest and freshest cheeses, such as yogurt cheese and chèvre, often have salt mixed into their curd. The mixing of the salt directly into the cheese helps to break up the curd and keep the texture as smooth as possible.

Some firm cheeses, such as cheddar, also have salt added directly to their curds. This pulls moisture out of the curds before they are pressed into a cheese—which helps cheddar develop its particular texture and flavor. Blue cheeses often have salt added to firm up the curds before forming to leave spaces between the curds, which leads to the development of blue veins. And feta is also salted before forming; this helps pull moisture from the curd before it is brined, giving it a firmer, more crumbly texture.

Cheeses whose curds are salted have the same quantity of salt added as if they were to be surface-salted. For example, 1 pound of chèvre would need about 1 tablespoon of salt, and 5 pounds of cheddar curd would need about 5 tablespoons.

CHAPTER SIX

Tools

The best tools to use in cheesemaking are those that are appropriate for you: Cheesemaking is adaptable to its tools, and there is no "best" tool to do a job. Materials and tools that are considered appropriate for making cheese in an industrial setting may not be the best for making cheese at home. And the tools that were used traditionally may not be suitable (or legal!) for commercial cheesemaking operations.

Alpine cheesemakers, for example, in the high pastures of the Alps fashioned curd cutters from wild-harvested pine saplings by cutting, bending, and tying their branches. The form and function of these traditional stirrers are mimicked in the form and function of modern stainless-steel curd cutters known as *spinos* (prim *pino*, Italian for pine tree) that are adapted to the conditions of industrial cheesemaking, whereas the real pine sapling versions would be outlawed!

Appropriate Technologies in Cheese

There are three different approaches to selecting the materials and tools to use in cheesemaking: the traditional, which is also the most natural; the industrial; and the improvisational. As a commercial cheesemaker, you may be mandated to adhere to the industrial standard by production regulations. But if you are a home cheesemaker, or an outlaw cheesemaker like me, you can take a traditional or improvisational approach: I use a mix of the three.

Traditional and Natural

When considering the natural in cheesemaking thus far, we have looked toward the traditional for inspiration, and traditional cheesemaking tools have much to teach us about a natural approach to handling cheese.

Historically, cheesemakers used wood or plant-based tools for making most of their cheeses. From the stirring paddles to the coopered cheese vats, from the willow cheese forms to the working and aging surfaces, cheeses would have been constantly exposed to wood. Such natural materials interacted with cheese in beneficial ways that the plastic and stainless-steel tools that have entirely replaced them in industrial facilities can never replicate.

Industrial

Most of the equipment used in industrial cheesemaking facilities is either stainless steel or plastic. Stainless steel is used for the vats and working surfaces and the majority of cheesemakers' tools, while plastic is used for draining mats, cheese forms, and cheese wrappings. Steel and plastic tools are desirable for commercial cheesemaking because they are mass producible and easily sterilized—two necessities for industrial cheese production. But for a smaller-scale and more natural cheesemaking, these two qualifications are of much lesser importance.

Industrial cheesemaking equipment can be quite costly, and investing in specific tools for cheesemaking can quickly add up. Though it's important to have dedicated tools for cheesemaking in larger commercial operations, home cheesemakers can source most tools from their own kitchen.

Improvisational

Household kitchen equipment can serve for most home cheesemaking needs. Pots of various sizes can be used as cheese vats; sieves and strainers can be used to strain the curd; and kitchen knives and wire whisks can be used as curd-cutting tools.

Other specific tools for cheesemaking, such as cheese forms, cheese presses, and draining racks, can easily be improvised out of common household objects. Plastic yogurt containers, for example, can be made into cheese forms; a pair of plastic buckets can be fashioned into a simple cheese press; and a sushi mat, placed atop a steel cooling rack balanced on a casserole dish, can serve as an improvised draining table.

Material Considerations

The materials you choose to make cheese with can affect the way a cheese evolves. If you wish to make cheese according to industrial philosophies, you should lean toward plastic and steel because they are more amenable to these methods; if you wish to make cheese using more traditional methods, choose wood and other natural materials that respond best to a natural cheesemaking. If you choose the improvisational, work with whatever you've got, and the cheese should turn out just fine!

The following are some considerations regarding the many different materials cheesemakers use.

Plastic

One of the many reasons I took up cheesemaking was that I didn't want to buy cheese wrapped in plastic. But as I soon learned, plastic is not just used in the final wrapping of cheeses; it permeates all aspects of cheese production.

Most industrial and artisanal cheeses made today will be in contact with plastic for the vast majority of their existence. Plastic is used in the cheese forms, in the cheese draining mats, in the wrappings that

protect the rinds of cheeses as they age, and ultimately in the plastic that cheeses are packaged in when sold. Should this permanent plastic exposure be a concern to cheesemakers and cheese eaters? I believe it ought to be.

Plastic use in cheesemaking is troubling for several reasons: Plastic is a pervasive pollutant, it's a nonrenewable petroleum product, and it can contaminate cheeses in contact with it.

Plastic waste from cheesemaking operations is significant. Even though much of the plastic used in cheesemaking is "recyclable," in actual practice plastic is rarely recovered from the waste stream. And what plastic is recovered is really only "down-cyclable": Recycling efforts never reclaim the original qualities of first-generation plastic, and recycled plastics tend to find their way into products that aren't ultimately recyclable, such as plastic "wood" and plastic fiber clothing. These second-generation plastic products have remarkably less durability, are generally not recyclable or recycled, and can create enormous pollution problems, both on land and in water.

Plastic is derived from oil, a dwindling resource whose use is dependent on an environmentally destructive petroleum industry. Even if the plastic used in cheesemaking were derived from plants (and it is generally not), the practices that go into the growing and processing of corn-based plastics are hardly benign.

Plastic may also leach chemicals used in its manufacture into cheeses that come in contact with it. Fat and plastic have an interchange between them: the milk fat in cheese, which can be seen being absorbed into plastic it comes in contact with, also absorbs fat-soluble components of plastic (you've probably tasted the plastic wrap in cheeses that have been stored in it too long). I generally recommend steering clear of plastic in cheesemaking, although for certain applications, such as cheese forms, its use is difficult to avoid.

As you'll see from the photos in this book, I use quite a lot of plastic in my cheesemaking. It is a most affordable, mass-producible, and easily

A selection of naturally made cheese forms.

Plastic cheese forms absorb milk fat and carotene from the cheese visible as a creamy hue on the form to the left. Do cheeses placed in such forms absorb elements of the plastic as well?

procurable material, and alternatives to it can be difficult to source.

Metal

Stainless steel is also a very common material in the modern cheesemaking operation. Cheese vats, kettles, and pots are all typically made of stainless steel, as are the vast majority of contemporary cheesemakers' tools.

Stainless steel is mandated for use for all the working surfaces of dairy processing facilities in most Western countries. Easy to keep clean, stainless steel used in cheesemaking must be scoured regularly with a mild acid (usually nitric and phosphoric acids, though you can also rinse regularly with vinegar) to prevent "milk stone," the buildup of milk minerals that compromise the ability to keep steel sterilized.

Copper is an alternative to stainless steel that is used in the making of many traditional European cheeses—and is even required for many cheeses such as Parmigiano Reggiano. The excellent

conductive qualities of copper make it particularly well suited to heating milk.

Heavy enameled cast-iron pots can help keep the milk warm much longer than stainless steel or copper and therefore work quite well for cheesemaking. Un-enameled cast iron, however, will give odd flavors to cheeses made in it and is generally not recommended.

Aluminum is to be avoided in cheesemaking for a number of reasons. First, it has a low heat capacity and does not make for good-quality pots or hand tools; and second, there is good reason to be concerned about exposure to aluminum in our food.

Wood

Sure, wood is difficult to clean and sterilize—but that didn't dissuade early cheesemakers from using it! In fact, the ability of wood to hold on to bacterial cultures made the making of their cheeses possible: All of the wooden tools that were used by traditional cheesemakers became imbued with the cultures of

Wooden surfaces absorb moisture, and culture, from cheeses.

cheesemaking and would have served as inoculants for the beneficial cultures that milk needs to become cheese. Several traditional European cheeses continue to be made in wooden vats, whose biofilms (aka slimes) are the cheese's primary source of starter and ripening cultures. The use of wood made possible the original culture of cheesemakers, who made their cheeses without an understanding of the nuances of microbiology.

I prefer wood for several other reasons: It's durable, it feels good in the hand, and it doesn't make a lot of noise, whereas metal tools rattle and bang and can disturb a peaceful cheesemaking session. (For this reason I also don't like to use metal tools in my cheesemaking classes.) And if you're using strong cheesemaking cultures, the risk of contamination from wooden tools is minimal; DVI cultures, however, with their inherent instabilities, demand the use of sanitized stainless tools.

Wood is commonly used to age fine wines and distilled spirits, much to their benefit. Similar improvements are realized when cheeses are made and ripened in contact with wood.

Pottery

As was recently discovered by archaeologists studying the remains of early human settlements around Europe, perforated pottery was the original cheese form. My own experimentations with ceramic cheese forms have yielded wonderful results!

Seldom used by cheesemakers in North America today because it is not easily sterilized, pottery is still used in the manufacture of many European cheeses. Some cheeses, such as Saint-Marcellin, are even left to age in small ceramic pots, which allow cheeses to ripen until their interior is molten in texture because the pot helps hold the cheese's shape.

Working Tools

Most likely you already have all the tools you need to start making cheese. No specific tools are needed for cheesemaking, except perhaps for cheese forms, though they, too, can be fashioned from common household objects. The following is a list of useful tools to have on hand when making cheese.

Hands

In many cases the finest tool for the job is simply the cheesemaker's hand. For gauging temperature, stirring curds, filling forms, and handling cheeses, no tool compares to the sensitivity and dexterity of a human hand.

The hand is the best tool for judging the firmness of the curd. In many cheesemaking recipes in this book, touch is used to tell if the curd is ready to proceed to the next stage of cheesemaking. The hand, with a little bit of training, can also serve as a thermometer for judging the temperature of an evolving cheese. Hands are handy for filling forms, as well as for flipping cheeses in the cheese cave. Wash your hands well before and occasionally during a cheesemaking session, but don't feel the need to sterilize them if you are healthy; the cultures on your skin do not interfere with the ecology of cheese and may, in fact, contribute beneficial microbes to their development.

Pots

A 1-gallon pot can be all you need to make small batches of smaller-sized cheeses such as chèvre, feta, and Camembert; and a 5-gallon pot can serve as a vat for making larger cheeses such as cheddar, Gouda, and Alpine cheeses. Good cheesemaking pots have heavy bottoms to ensure that milk isn't burned while it is heated; heavy pots also help to hold on to the warmth of the milk and prevent the developing cheeses from cooling off.

In larger operations, however, more specialized tools are needed. Round steam-jacketed kettles help in the making of cooked-curd cheeses such as Alpine cheeses. And rectangular cheesemaking vats make large-curd cheeses such as Camembert, cheddar, and Gouda much easier to manage. Stainless steel has all but replaced wooden cheese vats, which are today only permitted in the making of a handful of European cheeses.

Most of the basic tools needed for cheesemaking are standard kitchen staples; you likely only need to get some cheese forms to get started.

Hand Tools

Some good tools to have on hand for all cheesemaking include long-handled spoons or paddles, slotted spoons, and measuring cups and spoons for measuring milk and salt and other ingredients.

A long-bladed knife (preferably without a pointy tip that can scratch pots) can be handy for cutting curds for making large-curd cheeses such as Camembert, blue, and feta, while a wire whisk works for cutting the curds of Alpine cheeses extra-small. A ladle is useful to help transfer soft chèvre curd into cheesecloth and to skim milk of its cream.

Colanders, strainers, and stainless-steel bowls can help with the making of paneer, ricotta, and chèvre. Kitchen towels can come in handy for keeping pots of milk incubating during the culturing and renneting phases of cheesemaking, as well as for cleaning up spilled milk and whey.

The tools can be plastic, metal, or wood, depending on what you feel comfortable using, or on what local regulations demand.

Stove

A good stovetop is essential for making most types of cheese but is not necessary if you work with milk that comes straight from the animal and is still warm. Simple electric cooktops work just fine for heating milk, as do gas stoves.

Electric cooktops are great for heating milk, as well as for incubating it. The lowest temperature setting is too hot for incubating milk, but turning the element on for several seconds, then turning it off can provide a warm surface on which a developing cheese can stay warm for quite some time. The pilot lights on old and inefficient gas stoves are also the perfect place for keeping a pot of cheese warm.

Consider baking bread while making cheese: The warmth of the oven helps to create the perfect incubating spot atop the stove. The general consensus among cheesemakers is that bread baking and cheesemaking should not mix because of the risk of yeast contamination, but yeasts are essential for the development of most cheeses, and I've had no problems practicing the two arts side by side. Commercial cheesemakers, however, are mandated to make nothing but cheese in their certified dairies.

Cheesecloth

Cheesecloth is the material used to strain certain cheeses, such as chèvre and yogurt cheese; its fine weave holds back the curds but allows their whey to flow through and thus helps cheeses drain.

Invest in a good, strong, reusable cheesecloth. Cheesemakers traditionally turned to muslin, a thinly woven, unbleached cotton or linen cloth that works just great. After each use, muslin can be scrubbed and washed, then reused and reused until it's threadbare. Contemporary cheesemakers often prefer polyester or nylon cheesecloths that offer more durability. Many grades of natural and synthetic muslin can be purchased through cheesemaking suppliers or from your local dress supply shop.

The coarse cheesecloth available in grocery stores, however, is useless for straining cheese. Its weave is too open, its structure is too weak, and it is impossible to clean and reuse; its only appropriate use in cheesemaking is for bandaging cheddar, as will be explored in chapter 21. More considerations for cheesecloth will be provided in chapter 9, Yogurt Cheeses.

Cheese Forms

Cheese forms are perforated containers used by cheesemakers to give soft cheeses their shapes. Cheeses such as Brie, Camembert, Limburger, and small aged goat cheeses all get their forms because they are placed in appropriate-shaped cheese forms to drain.

Cheese forms are the most pervasively plasticized tools in the dairy. A visit to a local cheesemaker or a browse through a cheesemaking supply catalog will reveal how well used and well loved plastic cheese forms are. Plastic cheese forms work, and cheesemakers across the world rely on them. Easy to clean, cheap to produce, standard in size, they fit well into a modern and efficient cheesemaking operation.

But what of the traditional cheese forms? What did cheesemakers use to form their cheeses before the present era of petroleum and plastic? Soft cheeses were traditionally made of one of two types of forms: woven baskets or perforated ceramic pots. Plastic cheese forms are manufactured to replace these traditional forms, and you can see the similarities in two distinct types of forms. One type, made of fine plastic ribs, is made to mimic baskets made of woven reeds, whereas the other, a perforated plastic cup, is intended to replace ceramic cheese forms used in the making of lactic cheeses.

But something is missing in the plastic forms: an important interplay between the cheese and its form. The natural materials that cheesemakers once used to form their curd pulled moisture out of the cheeses and thus helped to preserve them. Cheeses that are in contact with plastic do not receive such benefits; rather, they dry more slowly and take on flavors and possibly chemicals from the plastic forms. Furthermore, every cheese made in mass-produced plastic forms will have distinctly the same shape as all the other cheeses made everywhere in the same forms. Handmade forms better reflect the character and individuality of handmade cheeses.

Homemade cheese forms can be made from a number of different materials. Improvised forms can be created from recycled plastic containers of appropriate sizes by punching through the plastic from the inside with a nail numerous times. Ceramic cheese forms can be made from pots punched though allover with small pieces of hollow grass stems, which leave behind small holes when the pottery is fired (the pots may be glazed or left unglazed). And small woven cheese forms can be fashioned from plants such as willow, reeds and ash; flat reed cheese baskets, such as those shown on page 65, are surprisingly easy to make.

Whatever material you choose to use for your forms—plastic, steel, ceramic, or reed—consider having a range of sizes and shapes for the range of sizes and shapes of cheeses you wish to make. And have several of each available, so that when you make your cheese, you have enough forms to fill.

Cheese Presses

Soft cheeses take the shape of their forms all on their own, but cheeses made to be harder, such as Alpine cheeses, cheddar, and paneer, will need to be pressed.

Most cheesemakers invest in mechanical cheese presses to achieve the pressure needed to unify the curds. These presses are often expensive and difficult to clean. Another method to press cheeses that's easier on the budget and simple to pull off is to repurpose two plastic containers of appropriate sizes into a homemade cheese press.

A simple small press for cheese made from 1 gallon of milk can be fashioned from two quart-sized (1 L) yogurt containers, one of which, the form, is skewered from the inside to allow whey to pass out, and the other, the press, is left whole to serve as follower atop the cheese in the form: When filled with warm whey, the follower becomes an excellent weight for pressing cheeses. A larger press, sufficient for pressing the cheese made from 5 gallons of milk, can be repurposed from used honey buckets, one punched through with a nail (from the inside) to serve as a form, the other left whole to serve as a press and a follower.

Though such passive presses do not apply as much pressure as mechanical presses, the warmth of the whey used as a weight encourages the cheese to stay warmer for longer, which allows it to be pressed firmer.

Draining Tables

A good draining table whisks away moisture from cheeses being drained on it.

Commercial cheesemakers lay down plastic draining mats atop sloped stainless-steel tables to facilitate whey drainage.

At home, you can set up a simple draining table by laying a bamboo sushi mat atop a steel cooling rack that's perched on a casserole dish. The sushi mat wicks moisture away from the cheese, the wire rack helps support the sushi mat, and the casserole dish catches all the draining whey. Cheeses are placed in their forms atop the draining table to drain; after salting, they can be laid directly on the

A simple cheese press can be fashioned from two plastic buckets, one punched through with holes, the other left whole.

draining table to dry for a day before being put away in the cheese cave to age.

Thermometers and Acid Meters

Two modern scientific instruments that industrial cheesemakers depend on are thermometers and acid meters. Neither, however, is necessary to make cheese.

Thermometers

Thermometers can be a useful tool for judging the right temperature of a pot of cheese, but by no means are they the only tool for that purpose. Consider that cheesemakers made excellent cheese for millennia before thermometers allowed us to measure temperature to the accuracy of a hundredth of a degree. I do not use one in my classes, just to show my students that cheese can be made just fine without them.

The hand is the thermometer of choice for making cheese naturally and traditionally, and with a little bit of training you can use it to determine with sufficient accuracy if the milk is at the right temperature for a specific stage of cheesemaking. Getting comfortable with your own sense of temperature can help you better control the warmth of the pot than is possible with scientific instruments— and it's not that hard to figure out. In fact, once you get accustomed to it, it's much quicker and simpler to use your hands to tell the temperature of a cheese than to use a thermometer.

There are only a few different temperatures that you will need to be able to gauge to make the cheeses in this book. And temperatures don't need to be exact: A variance of a few degrees won't drastically change a cheese. Here's some advice for judging important cheesemaking temperatures by hand.

At around 90°F (32°C), the temperature to which milk is warmed when cultured and renneted, if you put your wrist up to the pot, you will feel a gentle warmness—just as a parent judges the temperature of milk for their baby's bottle.

At about 110°F (43°C), the temperature to which Alpine cheeses are cooked, and the temperature at which yogurt is incubated, your finger will just barely be able to tolerate the heat. If you have to pull out your finger in pain after 10 seconds, the temperature is above 110°F. Though some may be inclined to believe that they have extremely high pain tolerance, the human body suffers damage at temperatures above 115°F (46°C), and our bodies are conditioned to tell us when it's too hot for our own good!

At the ideal temperatures for a cheese cave, less than 50°F (10°C), butter will be firm. Leaving butter in your cave alongside your cheeses can help you visualize if the temperature of the cave is too high. And so long as handling your cheeses doesn't leave your hands painfully cold, the cheese cave is likely at a good temperature.

If, however, you wish to pasteurize your milk, accurately measuring temperature with a thermometer is essential to ensure that the milk's cheesemaking ability isn't lost because of overheating.

Acid Meters

Acid testers are necessary tools for cheesemakers who choose to make cheese according to standard North American cheesemaking practices. They are not, however, a necessary investment for cheesemakers who use traditional methods; it is the use of freeze-dried DVI cultures that necessitate the use of acid-monitoring equipment.

Cheesemakers who make large batches of cheese with DVIs must assure themselves that their cultures are active by monitoring changes in their milk's pH. This is true for two reasons: There is no guarantee that the DVIs are alive when they are added to the vat, and DVI cultures are susceptible to bacteriophages—viruses that can kill freeze-dried cultures. There is much milk at stake in commercial cheesemaking operations, and cheesemakers use their acid meters to confirm that their cultures are working.

Using home-kept starter cultures such as kefir can eliminate the need for monitoring acidity. Knowing that the cheesemaking cultures are fresh and active by observing changes in the starter is all the assurance you need that your cheese will turn out as you expect. And traditional, biodiverse cheesemaking cultures have significantly better resistance to bacteriophage viruses and will not fail during a cheesemaking session.

Industrial cheesemakers are also reliant on acid meters to judge when their milk and their curds are ready for the next step of their processing. Cheesemaking recipes call for certain steps to be carried out at certain levels of acidity; however, there are ways to ensure that the curds are at the right stage of development without the use of acid meters, and this book will instruct you on how to recognize those different stages with that most important cheesemaking tool, your hand!

To Sterilize or Not to Sterilize

Sterilize, sterilize, sterilize is the mantra of industrial cheese production: Sterilize your milk, sterilize your tools, sterilize the entire cheesemaking environment. The North American Cheese Establishment has zero tolerance for foreign microorganisms.

Risk reduction plans (such as HACCP, an acronym that stands for "Hazard Analysis and Critical Control Points") in industrial cheesemaking facilities mandate sterilization of all tools, in part to keep each batch of cheese separate from the others in case foodborne illness arising from the production facility must be tracked to a specific batch. In facilities that produce thousands of pounds of cheese daily, with the milk from thousands of animals, such considerations may be important; but an analysis of a whole and natural cheesemaking operation in which batches of cheeses interact with one another as they are made and aged would be overwhelmed by so-called hazards, and the control points that would be needed to reduce the perceived risks would ruin the interactions that make cheeses evolve according to their natural way.

But it's not just food safety concerns that mandate biosecurity measures in cheese plants: The absolute need to sterilize all tools and equipment used in cheesemaking is a consequence of the use of DVI cultures. The two or three strains of bacterial cultures present in packaged DVIs are not stable communities of microorganisms. Each one is raised in a sterile laboratory setting and consequently needs a sterile setting to thrive. These cultures do not have the strength and durability of traditional cheesemaking cultures, like those found in raw milk or kefir; and if the cheesemaking operation is not perfectly sterile, rogue cultures will find their way into DVI-started cheeses. For best results manufacturers of DVIs even recommend using pasteurized or, at least, thermised milk. Raw milk's indigenous bacterial populations pose a threat to DVIs, whose viability is reduced by competition with raw milk microorganisms. The only way to get guaranteed results with DVIs is to pasteurize.

Keeping the cheesemaking operation perfectly sterile helps many cheesemakers eliminate the growth of wild microorganisms that can cause defects in cheeses made with DVIs. Indeed, all tools used in industrial cheesemaking operations, including a cheesemaker's hands, must be sterilized before use. To ensure the continued viability of DVI cultures, and to restrict the growth of wild microorganisms that find niches within the microbiologically unstable cheeses, cheesemakers have to be particularly careful about contamination.

Cheese is the perfect medium for the growth of bacteria and fungi, but if the right cultures get established in a cheese, their mere presence restricts the growth of unwanted microorganisms. Adding kefir culture to milk, for example, can help restrict the growth of wild *Penicillium roqueforti* and many other unwanted cultures because all of the ecological niches within the milk become occupied by kefir's many cultures.

Sterility is overkill: Clean is, I believe, good enough for cheesemaking. Creating the right conditions, not creating sterile conditions, can encourage the growth of beneficial microorganisms and limit unwanted cultures. By using strong, biodiverse cultures, and handling cheeses well, cheesemakers can approach their cheesemaking with less fear of foreign microorganisms. You can practice a clean, but not sanitized, cheesemaking, and your cheeses will not suffer from contamination if they are well tended.

·————·

CHAPTER SEVEN

The Cheese Cave

The ideal conditions for aging cheese can be found, year-round, underground. Cheeses prefer to be preserved in cool and humid spaces; caves, the original aging places, provided traditional cheese-makers with near-perfect humidity and temperature conditions for ripening their cheeses year-round.

The Cave

Sadly, few of us will have easy access to underground caves; if we want to age our cheeses we need to ripen them in a cavelike environment. To create the perfect cheese aging space, it is first important to understand what aging cheeses need: high humidity, low temperatures, free-flowing air . . . and a little bit of care.

The Importance of Humidity

Humidity promotes the proper development of aged cheeses. Most varieties of cheese ripen best at conditions of around 90 percent relative humidity.

High humidity helps cheeses preserve their integral moisture, necessary for the growth and development of their bacterial and fungal cultures. Too much humidity, however, can cause a cheese to be too wet, which can upset the microbiological balance within it.

Too little humidity can cause other problems: In a dry aging environment a cheese will dry out. And when a cheese dries, bacterial and fungal development slows, and the cheese stops ripening.

Low humidity can also cause cheeses to crack. Cheeses lose moisture from their surfaces; their rinds therefore dry out before the interiors. The drying rinds can shrink, which causes them to tighten around the cheese and makes them susceptible to splitting.

Some cheeses, such as feta stored in brines or Goudas sealed in wax, can be kept in environments of low humidity without consequence because their brines or rinds keep them moist. Aging cheeses in brines or wax can greatly simplify the aging process—but it limits the possibilities of a cheese.

The Importance of Temperature

The ideal temperature for aging most cheeses is 50°F (10°C) or less. Higher temperatures encourage cheeses to ripen more quickly, while lower temperatures slow the aging process to a crawl.

Temperature affects the growth rate of bacterial and fungal cultures within a cheese. Temperature, therefore, affects the speed at which cheeses age. A given cheese will ripen in much less time at 50°F (10°C) than it will at 40°F (4°C) because ripening cultures are much more active at higher temperatures.

At extremes of temperature, though, the balance of beneficial bacterial and fungal growth can also be upset, and a cheese may not ripen as intended. At too high a temperature, bacterial and fungal growth can become uncontrollable and can develop strong flavors in a cheese, while at too low a temperature certain cultures just will not grow.

High humidity can be preserved inside a cheese chamber: a large plastic container lined with bamboo.

Cheeses will ripen at low temperatures: Even in a refrigerator, bacterial and fungal cultures continue to grow, and a cheese will continue to age. Many cheeses are even placed into the cooler confines of a refrigerator (less than 40°F—that's 4°C) to finish their ripening. However, the balance of bacteria and fungi in a younger cheese can be upset in a refrigerator because some cultures grow better in cold conditions than others. For example, *Penicillium roqueforti* prefers cooler temperatures, whereas *Geotrichum candidum* prefers warmer. *P. roqueforti*–ripened blue cheeses therefore ripen better at cooler temperatures, and *Geotrichum*-ripened white-rinded cheeses prefer slightly warmer weather; thus, aging a white-rinded cheese at low temperatures can make it susceptible to blue fungal growth.

Overactive fungal development at high temperatures can also cause soft cheeses to liquefy excessively at the rind, which can make their rinds slip. And hard cheeses will leak their butterfat and become greasy in too warm an aging space.

The Importance of Air

Air movement is essential for the health of cheeses' rinds: In fact, the cheeses' need for fresh air is often described as a need to breathe.

The interiors of cheeses ripen just fine in the absence of air. Many contemporary cheesemakers even ripen their cheeses in vacuum-sealed plastic, creating an air-free environment: The cheeses ripen purely from the interior thanks to anaerobic bacterial cultures. Cheeses sealed in wax, or ripened in brines, also do not need air to thrive.

However, natural rind development depends on air: Aerobic bacterial and fungal species that grow on cheeses' rinds won't develop properly if a cheese can't breathe. The bacterial and fungal cultures that ripen white-rinded cheeses, blue cheeses, and washed-rind cheeses all need air to thrive.

But if air doesn't move around the surface of a cheese, a cheese's ripening can also go awry: Cheeses left without flipping or in contact with an impermeable surface for too long can become wet, which can change the way they age. Wrapping cheeses in plastic (without first creating a vacuum) causes humidity to build up on the rind to the point that an aging cheese becomes wet; one that's kept consistently wet can develop a washed-rind cheese ecology.

The Importance of Care

Cheeses must not be left neglected in their caves; a cheese will not develop as intended unless it is carefully tended to. We will explore how to care for aging cheeses later in this chapter (see Tending Aging Cheeses, page 80).

Small-Scale Cheese Caves

The act of aging cheese can be very rewarding and can help you evolve and expand your cheesemaking repertoire beyond fresh cheeses. Understandably, the idea of aging cheeses can prove intimidating to many beginning cheesemakers, but don't be discouraged—aging cheeses is simpler than you might think. In fact, you've probably aged cheeses in your own cheese cave without even realizing it.

If you've ever bought a Camembert and left it in its wrapping paper inside your refrigerator for a couple of weeks, you've successfully aged a cheese! Inside its wrapping (known as mini cheese caves in the cheese industry), the cheese preserves its integral humidity; and in the cool environment of your refrigerator, that Camembert continues ripening. If you were to cut into the cheese the very same day you bought it, you'd discover that it is disappointingly solid; however, if you leave it in its wrapping papers in your fridge for several weeks, the Camembert will ripen to perfection.

Building on your newly discovered experience in aging store-bought cheeses, you can create the same conditions for aging your own cheeses in your own cheese cave on a fairly small scale. Simple, small caves can be fashioned from large plastic containers, cheese wrapping papers, or even Mason jars, all of which can be placed in a cool environment to provide the right conditions for aging cheese.

Cheese Chambers

A large plastic container can serve as a perfect cheese aging chamber. Inside the sealed container, the integral humidity of the cheeses is preserved while air moves about them, allowing them to breathe. The cheeses are easily handled and flipped, and the container can be easily cleaned and maintained.

The cheese chamber can be placed in a cool space, and the container will then have the cool temperatures that cheeses need to age. In the fall, winter, and spring, the chamber can be placed into a cool room; in the summer months you can move it into the refrigerator to keep it cool.

As cheeses age, they respire moisture into their aging environment. If the cheese chamber becomes too wet, dry out the container with a clean rag, then put the cheeses back inside to age in a slightly drier environment. You can also line the chamber with a bamboo mat to keep the cheeses from contact with plastic and excess moisture.

Cheese Wrappings

Cheese wrapping papers, also known as mini cheese caves, can be used to create a microclimate for aging cheese in a cool but dry environment. Smaller, surface-ripened cheeses can be wrapped in wrapping papers that preserve the humidity that cheeses need to age, but also allow them to breathe.

Cheeses are usually wrapped in these papers only once they have aged for some time in a cheese cave or cheese chamber. If a cheese is wrapped too soon, the wrapping can trap too much moisture inside and negatively affect the cheese's ripening regime. Only after the ripening cultures have become firmly established on the rind of a cheese should a cheese be wrapped.

Wrappings allow commercial cheesemakers to package up and sell their cheeses once they've reached a certain stage of development, so that they don't take up precious aging space in their caves. Cheesemaking supply shops offer several different styles of cheese wrapping papers. Usually made of paraffin-waxed paper and breathable plastic, the wrappings prevent excess moisture from building up on the surfaces of aging cheeses while preserving the right level of humidity and allowing the cheese to breathe.

The use of these disposable wrappings has proliferated among artisan cheesemakers, and it's hard these days to find a Camembert ripened without them.

Alternatives to these expensive papers include butcher paper and waxed paper, both of which make excellent and affordable cheese wrappings. Even plastic wrap or plastic bags can work as cheese wrap; however, plastic can trap too much moisture on the surface of a cheese, and cheeses left in contact with it for too long can take on unwanted flavors. Traditionally, many types of cheeses were wrapped in leaves to help preserve their humidity (more on leaf wrapping in chapter 12, Aged Chèvre).

Aging in Mason Jars

Cheeses can also be aged in Mason jars! The jars' small size makes them perfect for creating a microclimate for individual cheeses: their lids can be lightly sealed to keep in just the right amount of humidity, and, depending on what space the Mason jars are stored in, the temperature of the aging environment can easily be controlled. The recipe for Mason Jar Marcellin in chapter 12 makes use of this handy glassware to create a perfect cheese aging microclimate.

Large-Scale Cheese Caves

There are many different ways to create a larger-scale cheese aging space. If you're interested in scaling up your cheesemaking operation, consider one of the following ideas for your cheese cave.

A Hacked Refrigerator

Refrigerators can make excellent cheese caves. But with operating temperatures of less than 40°F (4°C) and low levels of humidity, fridges do not make a perfect aging environment. To turn a fridge into a cave, you'll have to hack its environmental controls.

Devices that override a fridge's thermostat can be installed; these allow the refrigerator to operate at the ideal cheese cave temperature of 45°F (7°C). Thermostat controls, bought online and plugged

Butcher paper makes an excellent mini cave for aging cheeses.

into stand-up fridges or chest freezers, allow for more precise temperature regulation than refrigerators are typically designed for.

Refrigerators (unless they are older models with ice boxes that lack humidity controls) are also very dry spaces, and cheeses left exposed to the air of a refrigerator can quickly dry out. The humidity in a fridge-cave can be adjusted by leaving out pans of water or wet rags, both of which slowly evaporate moisture and thus improve the aging conditions. Cheese can also be aged inside cheese chambers or mini caves within the temperature-controlled refrigerator. This is one of the methods I use to ripen my cheeses: Many of the cheeses photographed in this book were aged in cheese chambers inside a hacked fridge.

A Wine Fridge

Cheeses can be aged, alongside your favorite wines, in a small wine fridge. With a wider range of temperature control than standard refrigerators, the climate of a wine fridge can be easily adjusted to suit the aging cheeses' needs. And depending on the humidity within, leaving a pan of water on the bottom of the wine fridge can help boost moisture levels.

The temperatures at which wines age match those that suit most cheeses. Historically the two were aged alongside one another in caverns and caves, making trips down to the cellar more efficient. Food safety regulations, however, now prohibit commercial cheesemakers from aging cheeses next to wines.

A Walk-In Cooler

A larger, walk-in cheese cave can be constructed by fitting a sealed and insulated room with a cooling unit and humidifier. Like hacked refrigerators, air-conditioning units can also be modified with thermostat controls that allow small spaces to be cooled to precise cheese cave temperatures.

Most commercial cheesemakers age their cheeses in such climate-controlled spaces. Industrial refrigeration units keep the temperatures cool, humidifiers keep the aging space humid, and fans keep air moving around the aging cheeses. Though they mimic traditional aging caves, unlike real caves, such commercial cheese aging facilities are energy hogs. The fans, humidifiers, and air-conditioning units all consume energy around the clock, and the electricity use of constructed caves is a major factor in the high costs of aged cheeses that must be ripened in the climate-controlled environment for many months.

Principles of natural building and passive solar design can be incorporated in cheese cave construction to reduce the energy demand of the aging space. Cheese caves can be built where deciduous trees cast shade on the building in the summer and help keep it cool. Caves can also be built into the north slope of hillsides and partly submerged in the soil to take advantage of the temperature- and humidity-moderating effects of the earth. The most effective way to reduce the energy demands of a cave, however, is to thoroughly bury it.

A Root Cellar

Though they're not as deep as caves, the temperatures in root cellars can be low enough for aging cheese throughout most of the year. In summertime, however, their temperatures are usually too high for aging many cheeses; only hard cheeses that preserve well at higher temperatures should be aged in root cellars in the summer.

Depending on the local climate, soil conditions, and design, root cellars may not be humid enough to serve as cheese aging spaces. Special considerations should be made, such as having exposed soil or installing humidifiers, to increase the relative humidity of the cellar; otherwise, cheeses will have to be aged in the cellar enclosed in cheese chambers or sealed in wax.

Tending Aging Cheeses

Cheeses left to age in their caves need regular care and attention. Much like a garden, aging cheeses respond well to careful tending and a little bit of love: ripening cheeses need to be flipped and kept dry, and the cave needs to be cleaned regularly to control cheese mites and flies.

Flipping Cheeses

Flipping cheeses regularly helps them ripen well. Flipping ensures that cheeses breathe, keeps them from sticking to their aging surfaces, and prevents them from sagging unevenly. Cheeses should be flipped every other day during the first few weeks of their development. But once they've established their ripening regime, they can be flipped once a week.

Cheeses may develop different ripening regimes on their rinds if they are left to ripen too long without flipping, particularly during their early days of aging when they're still impressionable. If one side of a cheese is left in contact with its ripening surface for too long, it may remain wet and develop a washed-rind ecology.

Cheeses that are left to ripen too long on one side may also become one with their ripening surfaces. A cheese that isn't flipped often enough may grow into the surface that supports it; then, when the cheese is eventually flipped, it may leave its rind behind!

Regular flipping also helps cheeses maintain an even shape. Cheeses are not a solid, but a gel; their semi-liquid nature makes their forms less than permanent. Cheeses, therefore, naturally sag over time, and if they're left to age on one side for too long, they will sag unevenly. Flipping ensures that they develop an even and attractive bulge.

Keeping Cheeses Dry

Cheeses must not get excessively wet while they are aging. They can become too wet if they are not properly dried before aging, if they are aged in an overly humid cave, or if they rest on impermeable surfaces.

Cheeses should be salted and air-dried before being put away in the cave. If a cheese is not properly dried before being aged, it could be exposed to conditions that are too wet. And if it's sitting on a wet surface, it likely will not ripen as intended.

Cheeses must be kept on breathable surfaces to age well. A cheese that sits on an impermeable surface such as plastic will sweat and become influenced by washed-rind cultures. If the surfaces on which cheeses rest are wet, they should be dried out.

Keeping the aging environment at the appropriate humidity helps to ensure that cheeses don't get too moist. Ensuring that there is free-flowing air in the cave also prevents the buildup of moisture.

Keeping Clean

The surfaces upon which cheeses age, be they wood, plastic, or steel, must be regularly cleaned, particularly during the early stages of the cheeses' development when they can be a bit messy.

I clean the aging surfaces of younger cheeses every two weeks and scrub the surfaces that support longer-aged cheeses about once a month. It is useful to have movable surfaces that can be taken out of the cave and given a good scrubbing with gentle soap and warm water and dried before being used again. A regular cleaning regime also helps control unwanted pests in the cave.

Controlling Mites

Cheese mites will inevitably find their way into your aging space. Commonly found in the farm environment, these little arachnids feed on grains, hay, and leaves, and when the opportunity presents itself, they'll also eat your cheese.

Cheese mites can sometimes be seen moving about on the rinds of aging cheeses, particularly harder, longer-aged types such as Alpine cheeses, cheddars, and Goudas. Not necessarily detrimental to the cheeses they eat, cheese mites are encouraged to thrive on certain aged cheeses such as Mimolette and are responsible for these cheeses' distinct rinds and sharp flavors.

If you don't want cheese mites on your cheeses, you'll have to work to prevent them from taking up residence in your cave. Regularly scrubbing the aging space can help control cheese mite populations. And the cheeses themselves also need regular preventive treatment to stop mites from becoming established on their rinds. Aging cheeses can have their rinds brushed regularly to keep mites from growing out of hand. Regularly rubbing rinds with oil, as with Alpine cheeses, and washing rinds with whey, as with washed-rind cheeses, also stop the growth of mites.

A cheese with an established infestation of mites can be difficult to treat. Mites burrow into the rinds of the cheeses and resist the treatments the outer rinds receive. Regularly treating cheeses as they age can help stop the mites before they become entrenched.

Grayish dust accumulating under aging cheeses is an indicator of the presence of cheese mites; a magnifying glass will confirm an infestation.

Controlling Flies

Cheese aging environments must be closed to prevent flies from entering. It's not just houseflies that are a concern, but fruit flies, too: The aging space must be protected to keep out even these tiny insects. For those that do get in, a simple fruit fly trap can be made by pouring some vinegar into a jar and adding just a drop of dish soap. Flies will be attracted to the vinegar but will drown in the jar because the soap changes the surface tension of the liquid.

If flies fly into the cave, they may lay their eggs on the rinds of the cheese. Their offspring feed off the protein-rich cheese and will eventually pupate and hatch into adults. If cheeses are left unprotected from houseflies and fruit flies, there may be unfortunate, or, according to some, serendipitous, consequences. An often talked-about but rarely sampled cheese, Casu Marzu, is made in Sardinia by the intentional introduction of cheese flies that turn the cheese into a wriggling, writhing delicacy.

I don't recommend allowing flies to raise their young in your cheeses; keep your cheese-aging environment free from flies at all costs!

The Seasonalities of Cheese

For every season there is a cheese. Like local fruits and vegetables that are only available at particular times of the year, certain cheeses are most suited to making and aging in certain seasons.

If you wish to make cheese without relying on artificial temperature controls, and rely entirely on the cooling effects of the earth, you'll find that the changing conditions within your cave make the space suitable for ripening certain cheeses at certain times of the year.

In fact, traditional cheesemakers only make certain types of cheeses at certain times of year, as milk quality and aging conditions vary considerably with the seasons. Interestingly, the milk quality of each season is perfectly matched to the types of cheeses that are best made in each season!

Spring

Cows, goats, and sheep feeding on lush spring growth produce abundant and delicious milk. But not much milk is available for cheesemakers in springtime as unweaned calves, kids, and lambs take their share, and milk is often only taken at one milking per day.

The short supply of milk, in combination with the cool weather, make spring an excellent time for making small, bloomy-rinded cheeses such as Camembert and Crottin. Perfect for picnics, these cheeses will ripen up just in time for summer.

Summer

Summer is the season for making massive hard cheeses. These cheeses make short work of the abundant milk that animals produce in summertime; and with the calves weaned (or slaughtered), and the grasses at their lushest, much milk flows. Summer milk is also the best for making hard cheeses. The quality of curds that result from pasture-fed animals makes it most responsive to the rigors of making firmer-fleshed cheese.

Summer conditions—high temperatures and numerous flies—are not conducive to the making

and ripening of the many different types of soft cheeses, but harder cheeses such as Alpine cheeses, cheddars, and Goudas all age well in warmer summer weather. These hard cheeses will ripen slowly over many months and can provide ample nourishment for leaner winter months.

Fall

Fall is the season for fungally ripened cheeses such as Gorgonzola and other bold blues. Fall is the season of mushrooms, and fungal cultures thrive in the cool and moist fall conditions. Take advantage of this seasonality to make your blue cheeses with the abundant and creamy milk that animals produce in the fall. These cheeses will ripen slowly over winter and take on fascinating flavors that can enliven the winter dining table.

Just as every vegetable has a season . . .

. . . there is a season for every cheese.

Winter

In the winter, animals fed forage do not produce milk that's high quality enough for many styles of cheesemaking. Cheesemakers know that winter is not the season for making massive cheeses such as Alpine cheese, cheddar, and Gouda, which require the milk to be as good as possible. If they make cheeses at all, they keep them small.

In temperate climates of the Northern and Southern Hemispheres, winter is traditionally a season of rest for livestock. Cows, goats, and sheep are bred in the fall to give birth in the spring, when lush spring grasses will help them produce the milk that nourishes their young.

Winter also gives cheesemakers a much-deserved rest. With all their work behind them, now is the time to enjoy the cheeses made in the summer and aged through the fall.

The winter solstice (and its associated holidays) are celebratory times that can be made even more celebratory with cheese!

CHAPTER EIGHT

Kefir

*K*efir (rhymes with *deer*) is a traditional yogurt culture from Central Asia. Often described as a fermented milk, kefir is a flavorful and slightly effervescent yogurt.

Kefir, though, is made somewhat differently from the yogurt most of us are familiar with. Like yogurt, kefir is made by adding culture to milk and encouraging a fermentation that sours the milk and thickens it into kefir. But unlike yogurt, when kefir is ready, the kefir culture is taken out!

Kefir is distinct among dairy cultures, as it forms colonies that have a touchable, feelable form. These colonies, known as kefir grains, are added to milk to ferment it into kefir; once the kefir is ready, the grains are strained out and reused over and over again.

Kefir is not a single bacterial culture but a community of diverse species of bacteria and fungi that live together in kefir grains. A Symbiotic Community Of Bacteria and Yeasts (SCOBY), kefir grains are known to contain dozens of bacterial and fungal cultures,

Kefir grains are the mother culture that makes kefir, a delicious fermented milk, a multicultural probiotic and the finest cheesemaking starter culture around.

each species playing a different role in the community, and all of the cultures together transform milk into kefir. Many of these cultures are closely related to those found in both raw milk and our digestive tracts, making kefir an excellent probiotic, as well as a superb source of culture for cheesemaking.

Kefir grains are easy to find (more information on procuring them is provided below), and if well taken care of, they will provide beneficial cultures for life.

Kefir Grains: A History

Kefir culture has been passed down from generation to generation for thousands of years. All kefir grains that exist today are descendants of the original grains discovered in Central Asia countless generations ago. Kefir's millennia-old cultural lineage puts 100-year-old sourdough starters to shame!

Kefir's culture is intertwined with the nomadic culture of Central Asia's high steppes. Stories from European travelers passing through the area many centuries ago describe the practice of keeping kefir as a continual fermentation of sheep or mare's milk inside a dried sheep's stomach, hung from the rafters of a yurt, from which kefir would be occasionally drunk.

Kefir likely helped these nomadic people drink the milk of their recently domesticated sheep, goats, and horses. Many residents of Central Asia are lactose-intolerant as adults, and kefir, with significantly less lactose than fresh milk as a result of its fermentation, would have been much more easily digested.

It is not known how kefir grains came to be—more is known about the origin of the universe than the origin of kefir! Given, however, that its microbiological profile is very similar to that of raw milk, and considering that the traditional practice of keeping kefir involved a continuous fermentation of raw milk, it is most probable that this multicultural organism evolved from the diverse cultures of raw milk.

All that can be surmised about its origins is that kefir's discoverers realized that the little grains that floated in their milk helped transform it into a tasty treat. Kefir grains, they recognized, could be added to their milk, and the milk would take on many interesting qualities: It would thicken up into a more luscious texture, and it would become slightly effervescent and . . . mildly alcoholic.

Kefir was likely kept, in part, because it transformed milk into a more celebratory beverage. Kefir's very name, which means "good feeling" in Turkish, reflects the slight intoxication one feels when drinking this fermented milk.

Kefir has been popular outside Central Asia for centuries. A favorite in Mongolia, it's thought to have been brought to the fringes of Europe by the Khans (along with Mongolian genes) in the tenth century CE. Popular in Eastern Europe and Russia for generations, its renown spread throughout Western Europe in the 1900s. It was with the counterculture movement of the 1960s that kefir found its way to North America, but it was not a household name until the current era of intestinal awareness. Kefir can now be found in almost any grocery store, having become more popular as people recognize its many health benefits.

The Probiotic Benefits of Kefir

We tend to think of ourselves in the singular, first-person sense. However, our self is just one of tens of thousands of species that make up our micro-biome. The beneficial bacteria that populate our skin, our mouths, and our digestive tracts, among other human habitats, outnumber our human cells 10:1! Summed all together, our commensal (meaning "eating at the same table") bacteria weigh as much as our brains, an appropriate organ to compare them to considering their critical role in our bodies' functions. Implicated in immune system function, dental health, skin health, and digestion, our commensal bacteria are necessary for our well-being.

Regular consumption of probiotic foods containing certain strains of beneficial bacteria has been shown to increase the numbers of commensal bacteria in our digestive system. Most commercially prepared probiotics, however, contain only two or three individual species of laboratory-raised bacteria, whose addition to the diverse microflora of the intestinal tract may be likened to a drop in a bucket. Kefir, on the other hand, with dozens of indigenous cultures, may well be the most beneficial of all probiotic dairy products.

Another advantage of kefir may be the high level of communication and cooperation among its different species. The individual cultures of kefir are nonpropagatable on their own; they need the company of other cultures within kefir grains to thrive. In our own digestive tract the interaction of the bacterial cultures plays an important role in our gut's health, and certain digestive tract ailments have been linked to noncooperation among digestive tract bacteria. Kefir may help to establish beneficial bacterial interrelationships in our bodies better than the isolated laboratory-raised bacterial strains that make up commercial probiotics.

Keeping Kefir Grains

Keeping kefir grains is much like keeping a pet: Kefir needs a little love and a bit of attention, and in return it will love you back. All that kefir grains need is a regular feeding of milk: Place kefir grains in milk, leave them to ferment at room temperature for 1 day, and they'll feed you back kefir.

Kefir grains prefer a daily feeding. Once the kefir has thickened, usually after about 24 hours of fermentation, the kefir is at its best. The kefir grains can then be strained out and used to ferment another batch of kefir. If left to ferment too long the kefir can develop too much acidity and a very strong, sour flavor and will begin to separate into curds and whey. If left to ferment at room temperature for over a week without a feeding, kefir can overferment, and the kefir grains may perish.

For home kefir-making, I recommend keeping between 1 teaspoon and 2 tablespoons (5–30 mL) of kefir grains. You can use that amount of culture to ferment between 1 cup and 2 quarts (240 mL–2 L)

of kefir. If you have less, the kefir grains will ferment too slowly, making the fermentation a bit more unpredictable; if you use more, the fermentation will happen more quickly and will be difficult to control.

Kefir grains, unlike us, aren't picky about which milk they are fed. Kefir grains can be fed raw milk or pasteurized milk, homogenized milk or unhomogenized milk, skimmed milk or full-fat milk; they don't have a preference for cows' milk, goats' milk, mare's milk, sheep's milk, or camel milk. However each different type of milk will yield a different-tasting and different-textured kefir. Just be sure that the milk is fresh: If the milk is older, it will have established cultures within it, which may get ahead of the kefir grains when the milk is fermented and can result in a strange brew.

Kefir culture, unlike yogurt, also doesn't need a specific temperature to thrive: There is no need to carefully incubate fermenting kefir, as the diverse bacterial cultures in kefir grains will ferment milk at a wide range of temperatures. Kefir's preferred temperature, though, is around 60 to 75°F (16–24°C), and kefir grains will produce excellent kefir in around 24 hours if left to ferment within this range. If the temperature is lower, kefir will take longer to ferment, and if the temperature is higher, the fermentation will happen more quickly.

Kefir grains will grow if they are well tended. If fed every day, kefir grains will double in size in one to two weeks. Kefir grains do not come out of kefir; only kefir grains beget kefir grains. And as the grains grow, bits of them can be split off into new kefir grains, which will grow all on their own. If you keep all of your growing kefir grains, very soon you will have too many, and your kefir will ferment too quickly and be hard to control. Most people only keep a certain amount of kefir culture, and eat, compost, or give away any excess grains.

Sourcing Kefir Culture

To make kefir, you'll need to source kefir grains. Fortunately, they are easy to find.

Kefir is the original "gift culture." Many kefir keepers originally received their grains as gifts and are happy to regift their excess grains to anyone who can give them a good home. If you cannot find anyone keeping kefir in your community, kefir culture can also be found on craigslist or Kijiji or purchased from online kefir breeders. I originally sourced my kefir grains from a fellow in Toronto who breeds them and sends them dried, in an envelope, to anyone who sends him $20 cash!

Kefir grains can also be found at many natural food stores. Several different culture companies sell dried grains, including Cultures for Health and GEM Cultures. But be wary of packaged kefir culture sold at grocery stores and cheesemaking supply shops: many brands of kefir are actually fake-kefir containing a collection of unstable laboratory-raised cultures, not propagatable kefir grains.

Fake-Kefir

An unfortunate proliferation of what I call fake-kefir has found its way into supermarkets across North America.

Naturally effervescent kefir is not easily sold in the supermarket, as its carbonation can cause plastic packaging to expand and burst! Most, if not all, kefirs available in grocery stores are therefore not made with true kefir grains, but prepared with a small selection of individual laboratory-raised bacterial strains that are chosen to represent kefir grains. These strains pale in comparison with the diversity of traditional kefir cultures and do not have any inherent effervescence.

The cultures of these fake-kefirs are not stable or nonpropagatable and likely do not have as significant a probiotic effect as true kefir. Fake-kefir will also not offer the same microbiological advantages as true kefir if it is used as a cheesemaking starter culture (it does not, for example, contain the *Geotrichum* fungus necessary for white-rind development).

That these fabricated fermented milks are labeled as kefir is misleading and takes advantage of consumers' trust. The name ought to be protected by a PDO (Protected Denomination of Origin—a certification that preserves the traditional manufacturing methods of many famous cheeses). This would ensure that all products sold by that name are made with real kefir grains, and thus carry all the true benefits of kefir.

Taking Time Off from Kefir

If you wish to take a break from kefir-making, your grains can be preserved for many months without care in the refrigerator—something you can't do with other pets.

Storing kefir grains in the cold conditions of a refrigerator slows them down, allowing them to be kept for many months without any care. To take time off from kefir, place the grains into a jar of fresh milk, and leave the sealed jar in the refrigerator. The culture can be kept for up to 3 months this way: Even though the milk would normally go bad in the refrigerator after several weeks, the kefir grains will preserve it by transforming it into kefir and, in the process, preserve themselves. When you're ready to make kefir again, take the grains out of the jar and place them in milk at room temperature. Because they are dormant, the refrigerated kefir grains will slowly ferment the first batch of kefir, which may not turn out quite as expected. But by the second or third feeding, the grains should be fermenting just fine.

Kefir grains can also be dried and will last for many months, even years. To dry kefir, place the grains on a cloth to air-dry for 3 or 4 days. The cloth will pull away moisture from the grains and leave them desiccated. When you're ready to make kefir again, the dried grains can be brought back to life by placing them in milk. Again, though the first batch of kefir may not turn out as expected because the grains are dormant, subsequent batches will turn out just fine.

Despite all the care and attention that kefir grains prefer, they can be neglected and still thrive. My kefir grains have seen a lot of mistreatment over the 10 years I've kept them: They've been microwaved, washed in chlorinated water, put in the blender, and recovered from the sink's drain. They've been turned into cheese, dried down and rehydrated numerous times, left untended at room temperature for far too

Mail-order kefir grains often come dried. They can be brought back to life by putting them in milk—just like dairy sea monkeys!

long, and even salvaged from the compost pile. Yet despite all of these mishaps, my kefir grains have remained loyal and continue to make delicious kefir.

If, ultimately, your kefir grains are killed (heat and an extreme amount of neglect are the two things kefir culture can't withstand), they will lose their firmness and disintegrate. If you wish to continue making kefir, you'll have to go out and source more kefir grains. Don't be too hard on yourself if this happens, for kefir grains are forgiving, and if you are generous and share your grains with friends, they'll happily share theirs if yours perish.

Material Considerations: Containers and Tools

If you want to make your kefir in a truly traditional way, consider making it in a bag fashioned from a sheep stomach! Hung in a sheep stomach from the rafters of a yurt, kefir made traditionally in Mongolia would be continuously fermented, with milk added as it was provided, and kefir taken out as needed.

I prefer to ferment my kefir in glass Mason jars—mostly because the glass allows me to observe its development. When it's ready, the kefir gels and begins to seep whey below. Mason jars come with standard lids that can be sealed tight, something that is beneficial if you wish to shake your kefir to develop its carbonation.

Kefir does not like to be in contact with metal for too long. Exposure to stainless steel is okay; however, the acidity of kefir may cause other metals to corrode. I make my kefir cheeses in stainless pots—sometimes exposing the kefir to metal for 2 days without causing any harm to my grains or seeing any off-flavors develop in my cheeses.

Plastic should be avoided in kefir-making, as the boundary between milk and plastic is not well defined. Plastic may leach chemicals into kefir; it also absorbs milk fat and can be very difficult to clean. Plastic containers are not generally recommended for use as fermentation vessels in any fermentative process, be it sourdough bread–baking, brewing, or yogurt-making, though many people do so without concern or complications.

Kefir is a strong culture that doesn't need coddling, as its antimicrobial properties restrict the growth of other bacteria and fungi. It's probably a good idea to wash containers and tools between use, but there is no need to sterilize them. Kefir can be kept in less-than-sterile containers, and the culture will not suffer—after all, the traditional method of making kefir was to make it continuously in a dried sheep stomach, without ever washing it out!

Kefir and Cheesemaking

Kefir may be the perfect cheesemaking culture. It's a biodiverse culture that is very easy to care for, simple to use, and nearly impossible to contaminate. The process of keeping kefir and using it for making cheese is similar to that for keeping sourdough starter for baking bread.

Every cheese in this book (and many, many more) can be made with kefir as a starter culture. It is a universal starter, containing both mesophilic and thermophilic bacteria that are adaptable to cheesemaking in any conditions. Kefir grains also serve as a source of bacteria for aging cheeses: because kefir contains bacterial species that feed on the products left behind by lactic acid bacteria. Kefir culture can therefore provide successions of ripening bacteria to any aged cheese. Cheeses made with kefir as a starter do not taste of kefir—their flavor is akin to traditionally made raw milk cheeses.

Kefir culture can establish raw-milk-like biodiversity in cheeses made with pasteurized milk and helps to ensure a strong and predictable fermentation in raw milk cheeses. Its microbiological biodiversity provides a broad range of cultures that are helpful for starting and aging many different styles of cheese. And its indigenous fungal cultures can help establish healthy populations of yeasts and fungi on the rinds of many types of aged cheeses.

Kefir contains many different species of yeast that can help set the stage for good bacterial and fungal growth on cheese rinds. It even contains fungal cultures such as *Geotrichum candidum* that contribute to the development of healthy white fungal coats on such surface-ripened cheeses as Camembert and Crottin. (See chapter 16, White-Rinded Cheeses, and chapter 12, Aged Chèvre Cheeses, for more information on encouraging *G.* growth.) Kefir's *G. candidum* can also help establish a strong washed-rind cheese ecology.

As a testament to the adaptability of this diverse culture, kefir can even be used as a starter for making a thick and luscious yogurt, the recipe for which is provided below. When given the right conditions, the appropriate cultures in kefir thrive, while the

unwanted cultures wither; provided with cooked milk, and a warm incubation environment, the right kefir cultures sour the milk and thicken it up into a beautiful, spoonable curd.

Kefir does not degrade as a culture. If regularly fed milk, the culture could theoretically last forever. Kefir grains are many thousands of years old and have been kept in less-than-sterile conditions for millennia. Despite exposure to many foreign bacteria and fungi, the diversity of cultures in raw milk, and even unclean conditions, kefir perseveres; it

Kefir culture is a diverse source of many beneficial cheesemaking cultures and can serve as a starter for making any kind of cheese. Conveniently, active kefir grains rise to the top of warm milk and can be easily removed from the curd once it has set.

does not require the coddling that freeze-dried cultures need to survive.

Kefir culture is simple to keep at home—it's perhaps the easiest of all dairy cultures to care for. Kefir grains are stable and do not degrade over time; kefir culture is resistant to the growth and proliferation of foreign bacteria and fungi; and kefir isn't just good to use as a culture for cheesemaking—there is an extra incentive to keep it: It's delicious to drink and offers a probiotic bonanza.

Needing a regular feeding, kefir is also a daily reminder of our responsibility to nourish ourselves and our communities, care for our culture, and give attention to our cheeses.

Adding Kefir Culture to Cheese

The best way to use kefir culture to start a cheese is to add the fermented kefir directly to the warm milk at the beginning of the cheesemaking process. One-quarter cup of kefir can be added per gallon of milk as a starter culture. For larger batches, 1 cup of kefir will suffice as a starter for 5 gallons of milk.

Just like when using sourdough starter to bake bread, the kefir should be active when used to start cheese. Kefir is best once it has fermented for a day and will remain active if kept in the refrigerator for up to 1 week. In the making of white-rinded cheeses I often let my kefir ferment for up to 2 days until it develops a thick coat of white surface fungus that can help to establish healthy white rinds.

Depending on its consistency the kefir can be difficult to mix into the milk; give your kefir a good shaking to break up the curd before you add it to the pot.

Kefir grains can be added along with the kefir, and conveniently, they will rise to the surface of the milk and break the curd when it sets, making them easy to retrieve. However, it's best to strain out the grains before you add kefir to a cheese—if a grain is left behind in the milk, it will end up in the cheese!

– RECIPE –
KEFIR

To make kefir, simply place kefir grains in milk. Left at room temperature, the grains will ferment the milk and thicken it into kefir in about 1 day. The

To make kefir, add kefir grains to fresh milk; leave the grains to ferment the milk for about 24 hours, until it thickens; then remove the grains from the kefir and start the process over again.

grains can then be strained out and placed once again in fresh milk, and the process repeated.

The making of kefir can take on a rhythmic nature: Because it takes kefir one full day to ferment, kefir-making can fit into a daily routine. If you feed kefir grains milk in the morning, by the next morning the kefir will be ready. Every day you can drink the previous day's kefir while you make preparations for the next day's.

I give my kefir grains as much milk as I wish to drink as kefir the next day. If I want to have a cup of kefir, I'll give my kefir grains a cup of milk. And I will keep only the appropriate amount of kefir grains to ferment that cup (240 mL) of milk—between 1 teaspoon and 1 tablespoon (5–15 mL). Any more and the kefir will ferment too quickly; any less and the kefir will ferment too slowly, wild microorganisms may dominate, and the kefir may taste odd. If you're making more kefir, add at least 1 tablespoon and no more than ¼ cup of grains per quart of milk (15–60 mL per litre).

I prefer to give my kefir grains a feeding every day to keep them active. But kefir can also be made in a larger quantity just once a week, and the kefir grains refrigerated in between feedings. However, kefir grains prefer to be fed regularly; keeping them in the fridge slows them down and makes their first fermentation a bit unpredictable.

Kefir can take on different consistencies and qualities depending on how it is handled. If left undisturbed while fermenting, it develops a lightly curdled texture. If given a regular shaking, it will have a more liquid texture and more developed carbonation. If you wish to have a kefir with a thicker yogurtlike texture—kefir doesn't quite compare to the texture of yogurt and can't be eaten the same way—consider making yogurt with kefir as a starter (see the yogurt recipe on page 94). The careful cooking of the milk and high-temperature fermentation involved in yogurt-making result in a "kefir" with a much thicker curd.

If you wish to have an effervescent kefir, take it for a walk! With every step you take the kefir gets shaken up, and the extra air that is mixed in encourages the growth of yeasts that help develop its carbonation. Too much shaking, however, can cause its cream to separate into butter (see chapter 23). An overshaken batch of kefir will have balls of butter floating on its surface.

Kefir is excellent on its own with a dash of salt added or a bit of honey, maple syrup, or fruit preserves mixed in preferably after fermentation. One of my favorite additions is fresh cherries, which, if added to the milk before it ferments, become delightfully effervescent when the kefir is ready.

INGREDIENTS

1 teaspoon (5 mL) active kefir grains
1 cup (240 mL) milk

EQUIPMENT

1 glass jar, with lid
Strainer

TIME FRAME

Around 24 hours

YIELD

1 cup (240 mL) of kefir

TECHNIQUE

Feed the kefir grains: Place kefir grains in a jar, and pour the milk over them. Cover the jar.
Let the kefir grains ferment the milk for 1 day: Leave the jar to ferment on the counter in the kitchen. There is no need to keep the fermenting milk either warm or cool—kefir grains like room temperature best. After a day, the kefir grains should thicken up the milk into kefir.
Strain out the kefir grains, and drink the kefir. You can use a steel or plastic strainer to strain the thickened kefir into a bowl below.
Rinse the kefir grains in cool water (if you so choose), and feed them again! The kefir can be eaten right away, or saved in the refrigerator for up to 1 week.

- RECIPE -
YOGURT

It takes yogurt to make yogurt. In the process of yogurt-making, milk is first cooked; once it's cooled, a small amount of yogurt starter culture is added to the cooked milk, and the inoculated milk is then incubated at a warm temperature until the yogurt sets. But though most yogurt makers source their yogurt starter either from packages of freeze-dried yogurt culture or with a bit of yogurt saved from a previous batch of yogurt-making, I make yogurt with kefir.

And though it may be an unconventional approach to yogurt-making (is there a recipe in this book that isn't unconventional?), using kefir as a yogurt starter works. That's because it's the cooking of milk prior to culturing, followed by a high-temperature culturing period, that transforms milk to yogurt, and not any specific culture that is added to the milk. Cooking the milk achieves several ends: First, it concentrates the milk, which results in a thicker curd as the yogurt ferments. Second, it denatures the albumin proteins in the milk into the curd, which also thickens the yogurt; and third, the high-temperature cooking and subsequent incubation reduces the oxygen in the milk, which encourages the appropriate types of fermentation that give yogurt its best qualities.

In the case of making yogurt with kefir, the high-temperature incubation encourages the development of *Lactobacillus* and *Streptococcus* bacteria, both endemic to kefir and both responsible for giving yogurt good flavor and texture. The low-oxygen conditions encouraged by the high-temperature incubation of the fermenting milk reduce the activity of kefir's yeast and fungal cultures, thus eliminating any effervescence that would normally develop.

In the struggle to attain the perfect yogurt curd, many yogurt makers pull out all the stops. Some believe that the only way to get a good thick yogurt

is to add powdered milk or milk protein concentrates, which, unfortunately, can have questionable origins. Others invest in expensive and unnecessary yogurt-making machines. Many replace their yogurt starter cultures frequently, especially those who use DVI yogurts, whose lab-raised bacteria are sensitive to contamination if reused. However, I've found that given that the milk is unprocessed, and the cultures are healthy and active, getting the right texture in yogurt is all about following the proper technique—specifically paying attention to the milk as it is cooked and incubated.

For yogurt to develop its best texture, its making demands the maker's full attention. The milk should be cooked slowly to a high enough temperature—185°F (85°C)—and for a long period of time, at least half an hour. The milk must be stirred nonstop as it is cooked, to encourage evaporation but also to ensure that it does not scorch on the bottom or form a skin on the top, both of which are coagulations of milk solids that take away from the thickness of the yogurt.

After cooking, you must cool the milk to the appropriate incubation temperature, stirring all the while to prevent a skin from forming. The cooled milk is inoculated with starter culture (the less you add, the better the result!), then the inoculated milk is incubated in a warm place to ensure it remains in the range of temperature at which yogurt sets best—between 100 and 110°F (that's 38 to 43°C). Once the milk has thickened, it has finished its transformation into yogurt and can be placed in the refrigerator to halt its fermentation, or left to ferment longer if you're looking for a more acidic yogurt.

I've found excellent results when carefully following this technique when using traditional yogurt cultures (many of which can be found in yogurt-making communities or from natural culture companies like Cultures for Health), or kefir. The advantages to using kefir as a yogurt starter, however, are that you need not bother keeping any other starters (something that deterred me from making yogurt with specific yogurt cultures) . . . and the

diversity of microorganisms within the culture make a remarkably flavorful yogurt. Furthermore, if the kefir is prepared with raw milk, the result will be a true raw milk yogurt.

INGREDIENTS

1 gallon (4 L) whole-fat milk, pasteurized or raw, preferably unhomogenized

¾ cup (180 mL) active kefir (strained kefir grains), prepared the day before, or yogurt

EQUIPMENT

1-gallon pot for cooking your milk (use either a heavy-bottomed pot or a double boiler to keep the milk from scorching as it cooks)

3 (1-quart or 1-L) Mason jars

Improvised incubator

TIME FRAME

1–2 hours of preparation; 4–8 hours of incubation

YIELD

Makes about 3 quarts (3 L) of yogurt

TECHNIQUE

Slowly warm the milk to 185°F (85°C) over medium heat. Stir the milk as it warms.

Cook the milk at 185°F for 30 minutes to 1 hour. Stir it nonstop. The longer the milk is cooked, the thicker the yogurt will be.

Cool the milk to 110°F (43°C): Take the pot of milk off the heat, and stir it until the temperature falls to that point. Don't have a thermometer? An age-old method of judging the perfect yogurt incubation temperature is to submerge your finger in the milk—if you have to pull it out in pain after 10 seconds, the temperature is just right!

Add the yogurt culture to the milk: Add ¼ cup (60 mL) of kefir or yogurt to each jar, then add an equal amount of the cooked milk, and mix the two. Fill the jars with the remaining milk, and put a lid on them.

To make yogurt, cook milk stirring all the while; add kefir or yogurt culture to the milk once it has cooled, and keep the milk warm until it sets.

Incubate at 100–110°F (39–43°C) for 4 to 12 hours, keeping the jars in a warm spot, such as an insulated cooler filled with warm water, or a warm oven with the light on. After 4 hours, watch the jars for visible signs of setting. Once the curd has set, your yogurt is ready though if left to ferment longer, it can develop its tanginess. Kept refrigerated, it will last a couple of weeks.

- RECIPE -
CRÈME FRAÎCHE

Crème fraîche (pronounced *krem fresh*—French for "fresh cream") is lightly fermented, thickened cream. By adding culture to full-fat cream, then allowing that cream to ferment, the cream slowly becomes acidic. Once it passes a certain acidity, the cream suddenly thickens into crème fraîche. If left to ferment longer, the thickened cream continues to sour and eventually becomes what North Americans know as sour cream.

Crème fraîche is used in a different way from sour cream. Because it is lightly fermented, and still has some sweetness, it's used the way most North Americans would use whipped cream. Best paired with fruit or dessert, crème fraîche is considered the perfect partner for strawberries. With more flavor than whipped cream, and less work to make (the bacteria thicken it up without any whipping!), crème fraîche can easily become a household staple.

Cream skimmed from raw milk will ferment naturally and deliciously into crème fraîche because of the indigenous raw milk bacteria that ferment its lactose into lactic acid. If cream has been pasteurized, bacteria need to be added back to the cream for it to ferment well. Most recipes for crème fraîche call for the use of commercial DVI cultures, or store-bought buttermilk, whose cultural origins are corporate as well. Kefir grains can be used as a sustainable culture for fermenting both pasteurized and raw cream into crème fraîche. Once the crème fraîche is thickened, the kefir grains float to the top and can be strained out and used again.

INGREDIENTS
1 tablespoon (15 mL) kefir grains, kefir, or whey
1 pint (480 mL) fresh, full-fat cream

EQUIPMENT
Jar
Strainer

TIME FRAME
24 hours

YIELD
Makes 1 pint (480 mL) crème fraîche

TECHNIQUE
Add kefir grains, kefir, or whey to fresh cream: Put the kefir grains in a jar and pour the cream over them or pour the kefir or whey into the cream. Seal the jar, and leave it out at room temperature.

Leave the cream to ferment, at room temperature, for at least 24 hours. Check up on it every so often to see if it has thickened, but do not disturb it too much.

Take out the kefir grains: Once the whole batch of cream has thickened (raw or unhomogenized cream may separate into thicker cream at the top and thinner at the bottom) the crème fraîche is ready. Pass the crème fraîche through a sieve to remove the kefir grains. Crème fraîche will keep, refrigerated, for up to 1 week.

Kefir Sodas

Kefir grains need not be limited to fermenting just dairy. Kefir can also ferment a wide variety of milk replacements and fruit juices. Soy milk, almond milk, apple juice, and grape juice are all suitable drinks to feed kefir grains. In fact, any sweet beverage can be fermented by kefir grains into a more interesting and probiotic drink!

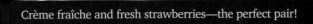
Crème fraîche and fresh strawberries—the perfect pair!

If kefir grains are added to grape juice, they'll ferment it into wine. If kefir grains are added to apple cider, they'll ferment it into hard cider. Place them in orange juice, and they'll ferment it into a kind of mimosa. Kefir grains can even be added to the sweet wort in beer brewing: The diversity of yeasts inherent in kefir (including *Saccharomyces* and *Brettanomyces*) can create very tasty ferments.

Saccharomyces cerevesiae, aka beer yeast, is an important member of the community of microorganisms within kefir grains. These yeasts feed on sugars, transforming them into alcohol and carbon dioxide. When fed milk, the yeasts in kefir grains contribute up to 1 percent alcohol in the kefir, and the carbon dioxide they generate helps to make the milk slightly effervescent. When fed sweet drinks with even higher sugar contents than milk, kefir grains can generate much higher alcoholic percentages.

The recipe below describes how to make kefir apple soda by fermenting fresh apple cider; however, any sweet drink can be used in place of apple juice. If you wish to make a harder drink, consider allowing the kefir to ferment the juice for longer— but just as in wine- and beermaking, be careful to limit the ferments' exposure to air, which can result in a vinegary tang.

Kefir grains, unfortunately, cannot be used to ferment drinks other than milk perpetually. Their community of microorganisms is dependent on milk and does not find all the nutrients it needs in milk replacements. If kefir grains are fed these alternatives over and over, the grains will eventually lose their vigor—but an occasional feeding of milk is all

Kefir grains can ferment fruit juices into effervescent and mildly alcoholic drinks.

that is needed to restore the microbiological balance the grains need to thrive. Water kefir, aka tibicos, is more suited to fermenting non-dairy liquids, so consider sourcing this other traditional culture if you prefer to feed your kefir grains only juice.

Interestingly, kefir grains take on the color of whatever drink they are fed. If you feed them grape juice, for example, they'll take on a purple hue and will keep that color even after they are kept in milk for several weeks—which makes them much easier to find!

- RECIPE -
KEFIR APPLE SODA

INGREDIENTS

1 tablespoon (15 mL) kefir grains

1 quart (1 L) fresh, sweet apple cider or other fruit juice *(be sure your juice doesn't have preservative agents in it such as potassium sorbate, which can kill the culture of kefir—along with the beneficial bacteria of your digestive tract!)*

EQUIPMENT

Narrow-necked bottle with lid

TIME FRAME

1 day–1 week

YIELD

Makes 1 quart (1 L) apple soda or hard cider

TECHNIQUE

Wash your kefir grains, rinsing off any cheesy residue in water.

Toss the kefir grains into the apple juice. Tighten the lid on the bottle, and leave it out to ferment at room temperature.

Allow the apple juice to ferment for between 1 and 7 days, removing the kefir grains after 1 or 2. Check up on the cider daily, releasing the pressure that builds up from the fermentation by unstopping, then restopping the bottle. (If too much pressure is allowed to build up, much of the cider will gush out when the bottle is opened.) Sample the brew occasionally, and enjoy when the taste, effervescence, and alcohol content are to your liking. The soda or cider can be kept refrigerated for 1 to 2 weeks.

CHAPTER NINE

Yogurt Cheeses

Yogurt cheese may be the world's most widely consumed cheese. Never heard of it? That's no surprise.

Variations on this cheese are known as labneh across the Middle East, *fromage frais* in France, and queso blanco in Spanish-speaking countries. But the English language doesn't even have a name for these cheeses; in English-speaking countries, if these cheeses can be found at all, they are labeled with a foreign tongue.

Why are these the world's most eaten cheeses? Because they are the cheeses that are the most easily made at home. And why are they not known in our culture? Because we lack a home cheesemaking culture. We are conditioned to fear for the safety of our dairy, so we leave our cheesemaking to professionals. And yet these cheeses are so easy to make, so safe, and so delicious that it's a wonder we don't make them ourselves.

To make yogurt cheeses you don't need a stove, you don't need rennet, and you don't need any special cheesemaking supplies. They are made by naturally thickening milk with bacterial cultures, then hanging that thickened milk to drain in cheesecloth until it becomes a soft, fresh cheese. That's it!

If you choose to make just one cheese, try one of these. The techniques for making yogurt cheeses are simple yet versatile, and quick but delicious. And those who wish to make cheese humane, or halal or kosher, by avoiding rennet can make these cheeses without compromising their values or beliefs.

The Process

Yogurt cheeses are made in a basic two-stage process: Milk is first thickened with the help of bacterial cultures, and the thickened milk is then hung to strain its whey and become cheese. Two additional steps can add flavor and longevity to these simple, fresh cheeses: salting, and adding herbs. To help explain all these stages and steps, here's a little more information on the process.

Curdling the Milk

When milk sours, it thickens. *Lactobacillus* bacterial cultures consume the lactose in milk and transform it into lactic acid. And as the milk becomes more acidic, its proteins, sensitive to acid, change their shape. The denatured proteins settle out of the milk and form a new structure that's known as curd. When that curd is then hung, it expels its whey and transforms itself into cheese.

Yogurt is but one of many curdled dairy products that can be hung to make cheese. Any fresh milk product, thickened with the aid of beneficial bacteria, can be hung in cheesecloth and turned into cheese. Cows' yogurt, goats' yogurt, sheep's yogurt, kefir, crème fraîche, and clabbered milk are all suitable candidates, and each one will make a very different cheese.

Yogurt, when hung in cheesecloth, becomes Dream Cheese, a beautifully textured fresh cheese reminiscent of cream cheese but with a much more complex flavor.

If goats' milk yogurt is hung in cheesecloth, it becomes chèvre. Chèvre made this way is tangier and creamier than the chèvre recipe presented in chapter 11 because it is made without the use of rennet. The curd retains more moisture as a result, which makes it softer and more acidic. Similarly, if you hang sheep's milk yogurt, you'll end up with a delicious cheese known as Brebis (pronounced *bre-bee* and French for "ewe").

Raw milk will curdle naturally when left at room temperature. The beneficial bacterial cultures present in the raw milk will sour and thicken the milk into a delicately curdled milk known as clabber. If clabber is hung in cheesecloth, it becomes a beautiful cheese that I call You Can't Do That with Pasteurized Milk Cheese.

Kefir can also be hung like yogurt; the result is a soft fresh cheese that many call kefir cheese, which is very similar in texture and flavor to hung clabber. Crème fraîche can also be hung in cheesecloth; when strained of its whey, it becomes a buttery cream cheese.

Straining the Curd

Straining the curd applies pressure to it and slowly presses out the whey. And as the whey drips out, the curd thickens into cheese. All told, it takes about 24 hours for curd to drain into a semi-firm cheese, though, depending on the desired consistency, the curd can be strained for longer or shorter periods.

The most effective way to strain curd is hanging. Curds can be hung by tying the curd-filled cheesecloth to a wooden spoon, then suspending the wooden spoon with its bundle of curds on the rim of a deep pot. Square-shaped cheesecloth can be tied to a wooden spoon by tying the four corners of the cloth together into a topknot, then sliding a wooden spoon between the knot and the cheese. Certain shaped cheesecloths, as discussed below, can make the tying and hanging much simpler.

The curd can be safely left at room temperature to drain into cheese. This traditional method of preserving dairy is safe and effective, and there need not be any concern about leaving the dairy out of the fridge. By getting the whey out of the curd and transforming it into cheese, its moisture content is reduced; as its moisture content is reduced, bacterial growth is slowed, and the cheese is thus preserved.

These cheeses need not be hung in the cold environment of the refrigerator to be preserved. In fact, if they are hung in the refrigerator, several problems may arise. Lowering the temperature of the curds causes the dripping of whey to slow down. As a result, it takes longer for the curd to drain. And because the cheeses are being exposed to the air of the refrigerator as they are draining, they absorb all sorts of noxious fridge odors: Onions, fish, and yesterday's leftovers are not flavors that go well with cheese!

Salting

Yogurt cheeses can be eaten directly after hanging, or lightly salted, then hung again to allow the salt to pull more moisture out. Salting helps remove moisture from the curd, which firms up the cheese and helps to preserve it by further slowing bacterial activity. If the cheeses are left unsalted, their high moisture content gives them a short shelf life: They will only keep for a few days if refrigerated. If salted, the cheeses will last for several weeks longer because of their low moisture content.

If the cheeses are well salted and dry enough, you can even preserve them by rolling them into balls and submerging them in olive oil, or by aging them into Blue Dream Cheese according to the recipes below.

Adding Herbs

Any herbs or spices, both fresh and dried, can be added to the cheeses. Most cheesemakers, though, add only dried herbs; adding fresh herbs adds extra moisture, which can reduce a cheese's shelf life. Nevertheless, fresh herbs can contribute much better flavors and colors than their dried counterparts—just be sure to eat the cheeses sooner.

The time to add herbs or spices to a cheese is at the end of the cheesemaking process. If flavorings are added while the cheeses are still being hung, much of their flavor will drain away with the whey. To add herbs or spices, simply mix them into the finished cheese, and allow the flavorings to meld for an hour or two before enjoying them.

Cheesecloth "Du's and Don'ts"

Get yourself a good piece of cheesecloth: anything but the despicable, disposable grocery-store-bought stuff, which has holes so wide that yogurt slips right through. To effectively use this lightly woven cheesecloth, you'd have to lay down five layers of it to hold back the curd; doing so, you use up almost the entire package. Furthermore, after only one use, this loosely woven material is impossible to clean without damaging the weave.

There are plenty of better materials that can be reused or repurposed as cheesecloth to make yogurt cheeses. Professional cheesemakers use muslin, a more strongly woven cloth made of cotton or nylon. Loosely woven silk scarves, polyester curtain sheers, bridal veils, and even nylon panty hose all work well as cheesecloth. I first made this style of cheese with a pillowcase and was thus inspired to name it Dream Cheese. However, cheeses hung in pillowcases tend to taste a bit like sleep, so be sure to sterilize and neutralize odors with baking soda and boiling water before using such materials to handle cheese.

Most highly recommended as cheesecloth, by me, is a du-rag. Du-rags, bandannas popular with rappers and gangsters, are made for cheesemaking! They have just the right weave for holding back the curd but still allow the whey to flow out, a head-shaped pocket that's just the right size for making a large batch of yogurt cheese, and two long tails that are perfect for hanging a cheese on a wooden spoon to drain. The long tails can also be used to squeeze the cheese, which shortens the hanging time and makes an extra-firm and round yogurt cheese that can be easily aged. They're simple to use, easy to clean, and cool, too: Du-rags make a gangsta cheese!

Du-rags make a gangsta cheese!

You can find du-rags at your local Jamaican barbershop: Just walk in, act natural, and ask for one. If you can't find them locally, some cheesemaking suppliers are beginning to carry them on my suggestion. Most, however, are made of synthetic materials; if you want a du-rag made of cotton, you can sew your own based on the simple pattern of a commercial du-rag, or knit your own out of natural fibers (see my Ravelry account—username: knitcheese— for a free pattern).

– RECIPE –
DREAM CHEESE

Dream Cheese is a cheese without borders: All around the world, people are making this cheese by leaving their yogurt to hang overnight to preserve it—the draining yogurt thickens up into cheese just in time for breakfast!

Though it tastes best if made with your own homemade yogurt made from good unhomogenized milk (chapter 8), Dream Cheese can also be made with commercial yogurt. However, be wary of which yogurt you choose for your cheesemaking.

Not every yogurt will work for making Dream Cheese. Many commercial brands of yogurt, particularly the lower-fat varieties, contain artificial thickeners such as pectin, cornstarch, or even gelatin. These thickening agents transform the yogurt into "yogurt pudding," which, when hung in cheesecloth to drain, will not release its whey. As a result, these unnatural yogurts won't turn into cheese; they'll just hang out in the cheesecloth and go sour and moldy! You can avoid this unfortunate scenario by carefully reading the ingredients list on your yogurt and choosing one made with only milk and bacterial cultures.

Wouldn't it be wonderful if blueberry yogurt could be hung into a blueberry cheese? Unfortunately, the fruit preparations added to many yogurts usually contain pectin or cornstarch, which prevent the yogurt from draining and thus from being preserved in the form of cheese. Instead, consider adding fruit preserves or—even better—fruit directly into the finished Dream Cheese.

Whole-fat yogurt hangs into a rich and creamy cheese. Low-fat yogurt works as well (it hangs into what's known as *Skyr* in Iceland), but because of its lower solids content, it makes less cheese. And for whatever reason—I think it's the lack of fat—it sticks to the roof of your mouth like peanut butter!

This recipe is not definitive, but rather up for interpretation. Depending on how long the curd is allowed to drain, the texture of this cheese can change dramatically. If you wish to have a creamier, quarklike cheese, let it hang for just 12 hours before salting. If you wish to have a firmer, *fromage frais*–like cheese, allow it to drain for the full 24 hours. Greek yogurt is another variation on this recipe; to make this lightly strained yogurt, let your yogurt drain for only 1 hour.

INGREDIENTS

1 quart (1 L) natural yogurt, homemade or store-bought; full-fat or low-fat; cow or goat, sheep or buffalo
1 teaspoon (5 mL) good salt (see chapter 5)

EQUIPMENT

Du-rag or other good cheesecloth
1 tablespoon (15 mL) baking soda
Large stainless or ceramic bowl
Wooden spoon
Deep pot

TIME FRAME

1–2 days

YIELD

Makes up to ½ pound (225 g) fresh cheese

TECHNIQUE

Clean and deodorize your cheesecloth: Place your cheesecloth into the bowl, and pour boiling water and baking soda over it to clean out any odors. Then rinse the bowl and the cheesecloth in cool water.

To make Dream Cheese, pour yogurt into cheesecloth; hang the yogurt to drain its whey; then salt the curd to firm it and preserve it.

Pour the yogurt into the cheesecloth: Line the bowl with the cheesecloth, and pour the yogurt into the cloth. Pull the corners of the cheesecloth together, and tie them into a topknot. Slide in a wooden spoon beneath the topknot. A du-rag, with its two long ties, makes the tying of the cheesecloth to the spoon much simpler.

Hang the yogurt: Suspend the wooden spoon with its load of cheese over a large pot. Cover with a clean kitchen towel to keep flies and other critters off your cheese.

Wait. Let the cheese drain for between 12 hours and one full day—the longer you leave it, the firmer it will be. Be sure that the cheese is suspended well above the level of the whey that is pooling in the pot below. If the cheese hangs too low, consider retying it so that it hangs out of contact with the whey. The cheese will lose up to three-quarters of its volume as it hangs.

Salt the cheese: Take the wooden spoon with its contents off its perch, and place it into a large bowl. Unwrap the cheese and investigate it; it should be nicely thickened and cream cheese–like. Pour in 1 teaspoon of good salt, and roughly mix it through the thickened curd with a spoon. Pull the sides of the cheesecloth back together, and tie it once again to the wooden spoon.

Let hang again for 4 more hours. The addition of salt draws moisture out of the cheese; hanging the cheese again is essential for preservation, as it allows that moisture to drain.

Enjoy the cheese as is, or mix in fresh or dried herbs. To mix in fresh or dried herbs, open up the cheesecloth and thoroughly mix in finely chopped herbs. Allow the flavors to meld by letting the herbed cheese rest in the refrigerator for at least 1 hour before eating.

Dream Cheese can be kept in the refrigerator for up to 1 week. It can also be preserved in olive oil (see the recipe for Shankleesh [Dream Cheese Preserved in Olive Oil], page 111), or aged into a rennet-free blue cheese (see the Blue Dream Cheese recipe, page 204).

– RECIPE –
YOU CAN'T DO THAT WITH PASTEURIZED MILK CHEESE

The reason I call this You Can't Do That with Pasteurized Milk Cheese is exactly as you might expect: If you follow this recipe with pasteurized milk in place of raw milk, you won't be able to eat the cheese that results!

Raw milk does not have as long a shelf life as pasteurized milk. But that fact conceals the reality that raw milk doesn't go bad like pasteurized milk: It goes different. The beneficial bacteria present in raw milk will sour and thicken it predictably and deliciously into what's known as clabber when it's left to ferment without a starter. If pasteurized milk is left to ferment in the same way, it will sour as a result of unwanted, wild microorganisms and will develop frightful flavors. Hanging rancid pasteurized milk into cheese will only concentrate the frightfulness; do not attempt to make this cheese with pasteurized milk!

If you are fortunate enough to have access to good raw milk, legally or otherwise, you can still make this delicious fresh cheese. It is made by first fermenting raw milk into clabber, then hanging that clabber in cheesecloth to drain into a soft cheese.

If you'd like to try the first cheese ever made, consider making this cheese. Our ancestors have likely been enjoying cheese made in this style for as long as dairy animals have been domesticated: Long before we ever discovered rennet, raw milk separated into curds all on its own through the activity of its native bacterial cultures. No doubt this curdled milk was strained to improve its flavor and preserve it. In fact, archaeological sites across Europe over 7,000 years old are littered with pieces of perforated pottery that did exactly that!

This cheese is a part of the common heritage for many North and South Americans, Europeans, Central Asians, and Africans. My own Eastern European family has a long history of making this traditional raw milk cheese: My great-grandmother Mirchka, who raised my grandmother and her seven siblings in rural Poland, made cheese in this style to sell in her small grocery in the village of Skawina.

Residents of Poland and other European countries where raw milk is widely available continue to safely make fresh cheese at home in this way. When immigrants from these regions arrive in North America and leave their grocery-store-bought pasteurized milk out to ferment, much as they used to with their grocery-store-bought raw milk, the disgust that inevitably ensues turns them off their traditional practice.

Substitutes for clabber that can be used to make a similar cheese include kefir, which when hung makes "kefir cheese," as well as cultured buttermilk (see chapter 23). Commercial cultured buttermilk, made from partly skimmed milk and added DVI cultures, however, will not make nearly as flavorful a cheese as fermented raw milk.

Making this cheese right requires raw milk; though obtaining raw milk may be legal where you live, producing a fresh (aged less than 60 days) raw milk cheese for sale is almost universally banned. I, however, have no concern breaking with convention and regulation to make a food that's an important part of my cultural heritage.

INGREDIENTS

1 quart (1 L) good raw milk (as explored in chapter 2)
1 teaspoon (5 mL) good salt (see chapter 5)

EQUIPMENT

Glass jar
Large bowl
Du-rag or other good cheesecloth
Wooden spoon
Large pot

TIME FRAME

2 days to make clabber; 1 additional day to hang the clabber into cheese

YIELD

Makes 6 ounces (170 g) fresh cheese

TECHNIQUE

Let raw milk sour into clabber: Leave fresh raw milk out at room temperature in a covered glass jar. Check on it every 12 hours or so. Once the raw milk has visibly thickened and begun to separate, you have clabber. It should take 1 to 2 days, depending on the time of year and the type of milk. If the raw milk is older, it will take less time to set, but the cheese that results may taste bitter because of psychotropic bacteria that thrive in the cold conditions of the refrigerator.

Hang the clabber: Line the large bowl with cheesecloth. Pour the clabber into the cheesecloth, tie the corners together, fix the package to a large wooden spoon, then suspend it over the large pot to let the whey drain for 24 hours. If you find that your clabber is too thin to successfully hang, there are several things to try: You can gently warm the clabber in a warm-water bath for an hour until it thickens, you can let your clabber ferment for several hours longer so that it separates more, or you can use a more finely woven cheesecloth.

Salt the cheese: Take the cheese off its perch, and place it in a bowl. Open up the cheesecloth, and mix the salt into the cheese.

Hang once again for 4 hours: Tie the cheese in its cheesecloth back to the wooden spoon, and suspend it once again. The addition of the salt will cause more whey to drain. After an additional 4 hours of hanging, the cheese will be ready to eat. You Can't Do That with Pasteurized Milk Cheese can be kept in the refrigerator for up to 2 weeks. It can also be preserved like Dream Cheese in Olive Oil (see the recipe on page 111) or aged into a rennet-free blue or white-rinded cheese (see chapters 17 and 16).

To make You Can't Do That with Pasteurized Milk Cheese, ferment raw milk into clabber; hang the clabber to drain its whey; and salt the curd to aid its preservation.

- RECIPE -
SHANKLEESH (DREAM CHEESE IN OLIVE OIL)

One delicious way of preserving Dream Cheese is to roll it into balls, dredge it in herbs, and submerge it in olive oil. This uniquely Mediterranean method of aging yogurt cheeses is popular from Greece to Israel and from Palestine to Iran.

To make shankleesh (this cheese's Arabic name), first make a firm Dream Cheese (see the basic recipe on page 104). The yogurt is hung into cheese, salted, then hung again for an additional day to ensure that it is extra-firm and dry. The firm cheese is then rolled into balls, dredged in a spice mix known as zaatar—a blend of thyme, sumac, and sesame found in most Mediterranean stores—then submerged in olive oil. The cheeses can thus be preserved for up to 8 months in a cool place.

Shankleesh is a very easy cheese to age and makes for a confidence-building first cheese-aging project. There is no need to create a humid aging environment to age your cheeses; just leave the jar in a cool space, and they will age submerged in the oil.

There's no need to worry about air being in the jar, nor about the need to sterilize the glass. These cheeses will preserve themselves below the olive oil, as their native bacteria, natural acidity, and low moisture content inhibit the growth of spoilage organisms and foodborne illnesses.

The oil will keep the cheeses from being exposed to air that would degrade them. As long as the cheeses are dry enough, they will stay submerged in the oil; and as long as they stay submerged they will be preserved. Two reasons cheeses might rise are that the aging environment is too warm and that high moisture content in the cheeses caused continued fermentation and the production of gases that give unwanted buoyancy. If the cheeses rise to the surface, they are not safe to eat.

INGREDIENTS

1 recipe Dream Cheese (page 104), made extra-firm and dry

4 ounces (113 g) zaatar (a mix of dried thyme, sesame, sumac, and oregano) or other dried spices

1 pint (480 mL) good olive oil or other oil

EQUIPMENT

1-quart (1-L) jar
Cool place for aging

TIME FRAME

1–8 months, or as long as you can keep from eating them

YIELD

Makes ½ pound (225 g) aged cheese— about 10 cheese balls

TECHNIQUE

Make Dream Cheese. Be sure to make your Dream Cheese extra-firm and dry—it should have a texture like modeling clay. To achieve this, salt your Dream Cheese well, and let it hang to dry after salting for an additional 24 hours. This firmness ensures that your cheese is as dry as possible before it is preserved in oil. Excess moisture in the cheese will cause spoilage.

Roll the Dream Cheese into cheese balls by hand. I make my cheese balls about 1 inch (2.5 cm) across. Wash your hands to remove any cheesy residue.

Dredge the Dream Cheese balls in dried herbs by rolling them, one by one, in a plateful of dried zaatar. The balls should be completely covered in herbs. Let the dredged cheeses rest for several minutes—this helps the dried herbs adhere to the cheese when submerged in oil.

Submerge the herbed cheese balls in olive oil: Fill up your jar halfway with olive oil. With the

To make shankleesh, first make a Dream Cheese that's extra-firm; then roll the cheese into balls and dredge in zaatar; finally, submerge the spiced cheese balls in olive oil.

aid of a spoon, lower the herbed cheese balls one by one into the jar. Once finished, be sure that the balls are all submerged below the surface of the oil, and add more oil if necessary. Finally, place a lid on the jar.

Age the cheese. This cheese keeps beautifully at low temperatures; it will certainly keep longer than you can keep yourself from eating it. It can be aged a minimum of 1 month and a maximum of 8 months. Ideally the shankleesh should be kept in a cool environment, like a cellar or an unfinished basement, but a refrigerator will work as well. If kept in a refrigerator, the olive oil will congeal: To safely remove the aged cheeses from the solid olive oil, first let the oil warm up to room temperature.

CHAPTER TEN

Paneer

A vegetarian cheese, paneer is greatly appreciated in India, where it holds a special place in the nation's vegetarian cuisine. Made without calf rennet, paneer emerges from milk in a different way from most other cheeses: A combination of heat and acid makes paneer happen.

To make paneer, milk is first brought to a boil and an acid such as lemon juice or vinegar is added. The milk, sensitive to acid at higher temperatures, curdles instantly, transforming into white, cloudlike curds and leaving behind a greenish-yellow whey. The curds are then strained from the whey, and, while still hot, pressed into a firm block of cheese.

Paneer, because it is made differently from most other cheeses, behaves differently, too. Paneer will not melt when heated, and therefore can be cut and cooked and will still hold its shape.

Paneer Biochemistry

It is because of the nature of milk proteins that paneer has such unique qualities. Casein, the most abundant milk protein, and the one that coagulates with the aid of rennet, is what most other cheeses are made of. It is albumin, however, the second most common protein in milk, that makes paneer distinct.

Albumin, found in egg whites but also present in milk, is a curious protein. A liquid at room temperature, albumin becomes a solid only when heated, which is the reason egg whites solidify when fried. Once albumin is heated, it denatures into a solid, and nothing can be done to turn it into a liquid again; a fried egg white just will not melt!

When cheeses are made at a lower temperature, the casein coagulates into cheese, but the albumin, still in its liquid form, remains with the whey (it is this albumin that gives whey the ability to make ricotta; see chapter 22, Whey Cheeses).

But when paneer is made at a higher temperature, both the casein and albumin proteins denature as they form into curds. As a result, paneer has a slightly higher yield than most cheeses because it incorporates more protein. And because the curds contain denatured albumin, paneer, like that fried egg white, will not melt.

As a result, interesting things can be done with paneer that just can't be done with other cheeses. For instance, paneer can be cooked in a sauce and will hold its shape and absorb the sauce's flavors like a cheesy sponge. Paneer can also be cut into cubes, stuck on a skewer, and barbecued. You can even make an excellent grilled cheese sandwich, with slices of paneer as the "bread," and a rich, meltable cheese such as Swiss on the inside—fried in butter, of course!

Considerations in Paneer Making

The following are considerations to help you make a better paneer.

Paneer, the perfect frying cheese, does not melt when heated.

The Question of Milk

Good milk isn't needed to make paneer. Milk that is overprocessed and won't make good rennet curd will still make fine paneer. Cow or goat, raw or pasteurized, whole or skimmed, even homogenized—just about any milk will do for paneer-making (except for UHT [ultra-heat-treated], which, because of its extreme processing, fails to make firm paneer). There is no need to go out and source unprocessed milk to make this recipe. In fact, if you go out of your way to get raw milk, making paneer pasteurizes that milk, eliminating many of its best qualities.

The relationship of rennet to milk is very intimate and organized and can be thrown off-balance if the milk is not of the best quality. The process of making paneer is a lot less organized than that for making a rennet cheese and so doesn't require as good a milk—making paneer by boiling milk and adding acid to it can be compared to scrambling eggs, while making a rennet cheese can be compared to the careful process of poaching an egg, which also requires that the eggs be as fresh as possible!

Your milk doesn't even need to be fresh. Any old milk will do, in fact, and you can even make excellent paneer out of milk that has passed its expiration date and give that milk a new lease on life! You can often get milk for free from your local grocer once it has passed its sell-by date, and so long as that milk does not have a sour taste it will still make excellent paneer.

Of course good milk will make much better paneer than old milk, or milk that is overly processed, or comes from industrial dairies. The more flavor a milk has, the more flavor the cheese will have, and the more nutrients that are in the milk (the more natural it is, the fresher it is, and the less processed it is), the firmer the curd will be, and the easier the paneer will be to make. The best paneer I've ever made came out of raw milk from pastured and well-loved cows.

Acidic Considerations

Depending on what type of acid you use, the flavor of the paneer can be drastically different. Traditional Indian paneer is made by adding yogurt, which, because it is only mildly acidic, needs to be added in greater volume to milk to separate the curd. However, when made with yogurt, paneer has the most natural, delicate flavor.

Vinegar is the acid I most recommend, and one that makes an excellent and firm paneer. Distilled white vinegar works, though many claim that it's only to be used as a cleaning product! Apple cider vinegar is the more natural, flavorful option, and one that gives paneer a slight apple-i-ness. Red and white wine vinegars work, too, as does balsamic vinegar, which gives a fine Italian flavor to paneer, and a slight purplish-brown hue.

If you wish to make a paneer that's more suited to dessert (paneer is excellent lightly fried in butter, with honey drizzled over it), lemon juice is a more suitable acid to use. But be sure to use freshly squeezed lemon juice, as paneer made of reconstituted lemon juice tastes of reconstituted lemon juice.

Citric acid is a food-grade acid that is often used to make commercial paneer. It has no place in a natural cheesemaking philosophy, though, as it is a highly processed ingredient derived from (most often) genetically modified corn. (See chapter 14, Pasta Filata Cheeses, for more information on citric acid.)

Keeping Paneer

Paneer is a cheese that cannot be aged. Because the process of making paneer involves pasteurizing the milk through cooking, the cheese that results is devoid of the bacterial and fungal cultures that help protect cheeses as they age. As such, paneer cannot safely be aged like other cheese made at lower temperatures.

Salting paneer can help to improve its keeping qualities, but it is not necessary if the cheese is to be eaten fresh. Paneer can be salted by adding salt into the curds prior to pressing. Another way to preserve paneer is to freeze it: This is one of the few cheeses that are not negatively affected by freezing.

– RECIPE –
PANEER

I learned how to make paneer at a *gurdwara* (a Sikh temple). The original community kitchens, gurdwaras open up their temples to the public and serve free vegetarian meals known as *langar* to anyone, regardless of gender, creed, or need, almost any day of the week. At the Golden Temple in Amritsar, India, the most holy Sikh temple, tens of thousands of pilgrims are served wholesome meals every single day.

If you haven't been to a gurdwara for a meal, I highly recommend it. It's an important cultural experience, and an excellent way to get to know your neighbors and enjoy a meal with folks off the street. If you don't want to accept a free meal, the temples will gladly accept donations, or your help in the kitchen.

Gurdwaras make phenomenal homemade Punjabi food, often featuring homemade paneer. When I learned that this temple I visited made its own cheese, I asked the community if I could volunteer in the kitchen and see how it was made. Expert cheesemakers, the Punjabis in the kitchen were very instructive and happy to share their skills. I later learned that many Punjabi households make their own paneer, even after immigrating to North America (you've probably seen them buying gallons and gallons of milk at the supermarket and wondered how they were going to drink it all). They should be an example for us all!

This is an adaptation of the gurdwara's recipe, scaled down from the 25 or so gallons (100 L) of milk that they transformed into cheese in their

kitchen! The 25 gallons of milk produced about 25 pounds (10 kg) of cheese, and all that warm cheese, sitting in the strainer, pressed itself firm. When making this recipe at home, you'll probably not be making as much, and you'll need to set up a cheese press to press your paneer firm.

Queso fresco, literally "fresh cheese" in Spanish, is a similarly made heat-acid cheese that's commonly consumed across Mexico and Latin America. Essentially paneer made on a different continent, the recipe for queso fresco is virtually identical to its Indian cousin.

INGREDIENTS

1 gallon (4 L) milk—and almost any milk will do!
½ cup (120 mL) vinegar (or 1 cup [240 mL] lemon juice, or ½ gallon [2 L] yogurt or kefir)

1 tablespoon (15 mL) salt (optional)

EQUIPMENT

2-gallon (8-L) capacity heavy-bottomed pot
Wooden spoon
Medium-sized wire strainer
Steel colander
Large bowl
Homemade cheese press—two matching yogurt containers, one with holes punched through from the inside with a skewer

TIME FRAME

2 hours

YIELD

Makes about 1½ pounds (700 g) cheese

Flavor paneer with any spice you like; here Mexican spices such as garlic, chipotle pepper, and coriander transform it into queso fresco.

To make paneer, pour vinegar into very hot milk; strain the curd that forms; and drain the curd either in a strainer or in a press.

TECHNIQUE

Bring the milk to a boil over medium-high heat. Be sure to stir the pot nonstop as the milk warms to prevent its scorching on the bottom; the more time you spend stirring, the less time you'll spend scouring! As well, stirring promotes presence of mind and keeps you focused on the milk, which may boil over if forgotten.

Let the milk rest by cooling it in its pot for a minute or two. Letting the milk settle will slow its movement and help ensure good curd formation.

Pour in the vinegar or lemon juice, and gently stir the pot once or twice to ensure an even mixing of the acid. Do not overstir; the paneer curds are sensitive when they're fresh and can break apart if overhandled. Watch as the curds separate from the whey . . .

Let the curds settle for 5 minutes. As they cool, the curds will continue to come together. As they become firm, they will be more easily strained from the pot.

Carefully strain the curds: With a wire-mesh strainer, scoop out the curds from the pot, and place them to drain in a colander resting atop a bowl that will catch the warm whey. Pouring the whole pot through the colander is not recommended, as the violent mixing that results can make it difficult for the cheese to drain.

Add spices or salt (optional). If you wish to flavor your paneer or queso fresco, consider adding various herbs or spices to the curds before they are pressed. Now is also the best time to add salt.

Press the curds (optional): Transfer the paneer curds from the colander into a form while they are still warm, and place the cheese-filled form atop a draining rack. Fill up the follower with hot whey, and place atop the form to press the curds firm. The paneer is ready as soon as the curd has cooled. It can be taken out of the form and used right away, or refrigerated in a covered container for up to 1 week. Paneer, unlike other cheeses, can also be frozen.

CHAPTER ELEVEN

Chèvre

If you've got good goats' milk, you'll be shocked at how little effort is needed to transform it into chèvre. Short for the *fromage de chèvre* ("goats' cheese" in French) chèvre is a fine fresh cheese that takes about a day and a half to make. But for most of that time, chèvre makes itself; all you really need to do is give the goats' milk a little guidance.

An excellent introduction to rennet cheeses, the process of chèvre-making is decidedly simple, involving eight steps that are undemanding, are easy to follow, and yield great rewards. Known as a lactic cheese, chèvre fits between simple lactic-acid-set yogurt cheeses and more complex rennet-set cheeses, containing elements of both styles of cheesemaking.

Interestingly, this lactic cheesemaking method is uniquely suited to goats' milk—a perfect marriage of medium and method. Goats' milk responds just right to the chèvre-making method. The process involves a long, slow fermentation that brings out the best flavors in goats' milk; a goats' cheese made more quickly, or a lactic cheese made with cows' milk, just doesn't taste the same.

Chèvre's Story

Chèvre (pronounced *shev*) is the cheese of choice in central France, where goats are the predominant milking animals. The foundation of a long-standing tradition of farmstead cheesemaking, chèvre is made by many *fermiers* and *paysans* from the milk of their small herds of goats in countless small dairies across the country.

Chèvre evolved in frugal farming households of the sort that continue to make it today. It is a cheese that's very economical, in both time and ingredients; made on the family farm, where there are many chores to take care of and livestock to feed, a cheese that didn't need much attention or many costly ingredients fit right in.

Goats: Activist Animals

Though they are often called "poor persons' cows," I prefer to see goats as activist animals. The milking animal of choice for homesteaders, back-to-the-landers, and peasant farmers the world over, goats are an antidote to Big Dairy.

Goats are a belligerent species that have rejected the rigorous production regime thrust upon their bovine cousins. Unlike cows, who contentedly chew their cud in confinement and produce enormous quantities of milk year-round, goats refuse to be cogs in the machine of industrialized dairying.

Expert escape artists who resist being kept indoors and prefer a diverse diet of browse, goats do not tolerate being kept in close quarters. Their refusal to be bred at any time of the year, their limit of 1 gallon of milk a day (modern Holstein cattle will often give over 10), and their milk that only stubbornly gives up its cream has stopped them from becoming commodities like cows.

Goats: activist animals that refuse to submit to industrial dairy practices.

Goats have not been the subject of intensive breeding campaigns to optimize production on a grain-based diet. There are no hormones developed explicitly to improve their production. Not politicized and controlled like cows (many regulations that control the flow of cows' milk may not apply to goats'), they are an accessible and affordable "fringe" livestock. And for counterculture cheesemaking, goats provide excellent milk that responds well to a simpler cheesemaking method.

The Chèvre Method

To make chèvre, goats' milk is first warmed to a comfortable temperature. Next, starter culture and a small dose of rennet are added to the milk, which is then left to ferment unattended for around 24 hours. The soft and silky curd that forms is scooped out, strained into a cheesecloth-lined colander, and left to drain its whey for about 6 hours. Once the curd has thickened, some salt is added, and the cheese is strained for 1 more hour to help the salt drain more whey. That's it.

Chèvre is a cheese that's particularly suited to home cheesemaking, and one that's easy to get into the habit of making. For most of its making the cheese just sits and takes care of itself. There are no periods of time when you really need to focus on the cheese, no need for careful incubating, and no delicate stirring.

As well, the schedule of its making is slow and forgiving. If you can't get to the curd in time, that's okay—whenever you can make time for it, chèvre is happy to oblige. If you leave the curd to ferment for 12 hours, 24 hours, or even 2 days, that's fine with chèvre. If you leave the cheese to drain for 6 hours, 8 hours, or even overnight, that's fine with chèvre, too! A truly French cheese, chèvre is the most *laissez-faire* of all cheeses.

Good Goats' Milk

Raw goats' milk makes the finest chèvre. A completely different cheese from flat-tasting pasteurized milk chèvre, raw chèvre is nuanced and flavorful.

The diversity of beneficial microorganisms in raw goats' milk is the definitive difference between raw chèvre and pasteurized chèvre, which is usually made with but one or two added strains of freeze-dried bacterial cultures. Though illegal to sell in most jurisdictions (it is not a cheese that can be aged the requisite 60-day minimum), raw chèvre is not illegal to make, and definitely worth its very minimal risk to eat.

You can taste the biodiversity in raw chèvre. Similar to fine wine, when eating raw chèvre you'll want to savor the flavor in your mouth as long as possible before you cherish the next bite. It is the many different microorganisms in the raw milk that contribute the many layers of taste in raw chèvre, each strain of bacteria fermenting the milk in a different manner, each one producing unique compounds that add complexity to a cheese.

If you cannot source raw goats' milk, seek out a goats' milk that's as fresh and unprocessed as possible; no matter how you handle it, overprocessed goats' milk just won't form good chèvre curds. And unfortunately, as is the case with cows' milk, most supermarket goats' milk is far too processed to make good cheese (even though goats' milk does not separate its cream to the same extent as cows' milk, most grocery-store goats' milk is still homogenized).

Goats' milk is often described as having a "goaty" taste; however, not all goats' milk can be accused of that. The goatiness of goats' milk is not inherent in the milk of this animal; it is a flavor that is imparted to the milk in a number of ways, all of which can be controlled through good dairying practices.

Keeping the buck, or billy goat, away from the milking does can greatly reduce the goatiness of goats' milk: Billy goats cultivate a certain stink about them through certain unmentionable habits (they urinate on their beards!), and their unmistakable aromas can be picked up by the does and passed to their milk. Ensuring that the goats have adequate fresh pasture can also reduce the goatiness in their milk, as can keeping their barns and milking parlor clean.

Whether it's raw or pasteurized, be sure your goats' milk is fresh. Goats' milk can also take on unwanted odors if it sits refrigerated for many days. Older pasteurized goats' milk and poorer-quality raw milks may also harbor unwanted coliform bacteria. These bacteria may cause your chèvre curd to float (the only thing that ever goes awry with this process)—and may make a cheese that is unsafe to eat. If you find that your chèvre curd floats after fermentation, consider pasteurizing it before making your next batch of cheese, but before you resort to such harsh treatment, be sure your milk is as fresh as possible, and seek out the possible source of contamination.

Culturing Chèvre

Traditional chèvre makers cultured their goats' milk with backslopped whey or through what many today would consider "unhygienic" practices that unintentionally cultivated beneficial cultures in their milk. If you're using raw goats' milk, you can continue the tradition of using leftover whey from a previous chèvre-making session as a starter: pasteurized milk, however, does not provide a stable source of culture for backslopping.

Commercial chèvre makers using pasteurized goats' milk start their cheeses with mesophilic freeze-dried DVIs containing but a few strains of laboratory-raised bacteria. Commercial chèvre's flat flavor reflects the sterile, laboratory-like conditions in which it's made.

If you're using pasteurized goats' milk, kefir can better restore the microbiological diversity needed to develop a more flavorful raw-milk-like chèvre. Ripe kefir can be added directly to the pasteurized goats' milk to give it the culture it needs to develop its most distinguished flavors; adding kefir as a starter can also ensure a predictable fermentation in a raw milk chèvre.

Soft and creamy chèvre curd made with a light dose of rennet and kefir culture, visible floating atop the curd in the top right of the pot.

Soft at Heart

Chèvre, just like a milking doe, is a soft and friendly cheese. It is mainly the dose of rennet added to the milk that makes it so.

One of the most important differences between soft and hard cheeses is the amount of rennet used. Semi-firm cheeses such as Camembert and blue cheese have a regular dose of rennet; firm cheeses, such as Alpine cheeses, use a double dose of rennet; and chèvre, the softest rennet cheese, involves adding just enough rennet to make the curd set.

Chèvre is a soft, smooth cheese; using a full dose of rennet will yield a chunky result. The rennet dose greatly affects the texture of the chèvre, and a greater dose of rennet will make a cheese firmer. However, it's not just the rennet dose that matters but also the quality of the goats' milk. Well-raised, pastured raw goats' milk will form a very nice curd with a tiny, almost homeopathic dose of rennet: just one-quarter of a regular dose. However, pasteurized goats' milk, or milk from goats that do not get much browse, needs a slightly higher dose of rennet to form good curds. It may take a few tries to get the rennet dose right: It's difficult to measure such a small rennet dosage, especially when making small batches of chèvre.

Slow Fermentation

Goats' milk responds beautifully to the slow fermentation it receives during chèvre-making. The unique fats in goats' milk, when subjected to a long fermentation period, break down into particular fatty acids that give chèvre much more flavor than a similar cheese made with cows' milk.

To encourage a slow fermentation, chèvre-makers use lower temperatures than are employed for other rennet cheeses. To begin, the milk is warmed to udder temperature to ensure rennet coagulation. Once the rennet and culture are added to the milk, the fermenting milk is allowed to cool to room temperature, thus slowing the fermentation.

Let the pot of milk ripen at room temperature for a day or longer: The longer you let the goats' milk ferment, the better your chèvre will taste. Don't wait too long, though, as a film of white fungus that originates in raw milk or kefir may begin to grow atop the whey in the pot after a few days. If, however, you are making certain types of fungally aged chèvres (see chapter 12), the fungal development that occurs after several days' fermentation can help to ensure an even growth of the cheese's fungal coat!

– RECIPE –
CHÈVRE

The cultural circumstances within which chèvre evolved make the production of this cheese ideally suited to our modern times. With the many distractions and diversions in our lives, it is often difficult to find dedicated time for cheesemaking; chèvre's simplicity helps it find a place in our daily rhythms.

Cows' milk can be used in this recipe in place of goats' milk: the soft and creamy curd that results is firmer than yogurt cheese and is sometimes called cream cheese, *fromage frais*, or Neufchâtel, though that final name is an American bastardization of a very different bloomy-rinded French cheese. The long fermentation of the cows' milk allows its cream to rise, creating a beautiful layer of creamy curd atop the whiter curd below.

Chèvre is excellent on its own but also serves as a delicious canvas for adding many other herbs, spices, and flavors. Roasted or raw garlic, cracked pepper, preserved lemons, even fruit preserves all pair well with chèvre. But be sure to add them at the end of the cheesemaking process, when the cheese is salted and drained; if the flavorings are added too soon, their flavor will flow away with the whey.

Chèvre is generally eaten fresh in North America, so it is a little-known fact that it can also be aged! Chèvre is the foundation of an entire class of aged cheeses that start as this fresh cheese (see chapter 12: Aged Chevre Cheeses).

CHÈVRE

INGREDIENTS

1 gallon (4 L) good goats' milk

¼ cup (60 mL) kefir or active whey

¼ dose rennet (I use less than ¹⁄₁₆ tablet WalcoRen calf's rennet for 1 gallon milk)

1 tablespoon (15 mL) good salt (see chapter 5)

EQUIPMENT

1-gallon (4-L) capacity heavy-bottomed pot

Wooden spoon

Ladle

Du-rag or other good cheesecloth

Steel colander

Large bowl

TIME FRAME

30 minutes to make; 2 days total

YIELD

Makes about 1½ pounds (700 g) chèvre

TECHNIQUE

Warm the goats' milk to around 90°F (32°C) on a low heat, stirring occasionally to keep it from scorching.

Stir in a cheesemaking starter culture: Pour in the kefir or whey and mix it in thoroughly.

Stir in a small amount of rennet: Dissolve the quarter dose of rennet in ¼ cup (60 mL) cold water. Mix it into the warm milk gently but thoroughly.

Leave at room temperature, covered, for 24 hours. After the long fermentation period, the curd will shrink and sink to the bottom of the pot.

Ladle the curds into a cheesecloth-lined colander perched over a bowl to catch the whey. Tie the cheesecloth into a bag, and simply leave it in the colander to drain.

Drain for at least 6 hours, at room temperature. Cover with a clean towel if need be to keep flies from landing on it. Be sure that the curds are well suspended above the level of the whey.

Salt the curds: Open up the cheesecloth bag and sprinkle 1 tablespoon (15 mL) salt over the surface of the cheese. With a wooden spoon, mix the salt into the cheese thoroughly.

Tie up the cheesecloth bag, and let the salted curds drain for another hour or two. Once the cheese feels quite dry, it's ready to eat, or have herbs or spices added to it.

Keep chèvre in the refrigerator if you don't eat it right away. It will keep for at least 2 weeks.

To make chèvre, ferment goats' milk with rennet until it yields a soft curd; hang the curd to drain its whey; and salt the cheese to preserve it.

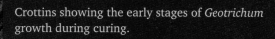

Crottins showing the early stages of *Geotrichum* growth during curing.

CHAPTER TWELVE

Aged Chèvre Cheeses

Goats' milk cheeses are the essence of goat. And while chèvre is like a gentle milking doe, aged chèvre is more like a billy goat: sharp and feisty, and leaving an odd taste in your mouth.

This class of aged cheeses takes on some of the most fascinating flavors as they ripen. But aged chèvre doesn't just have flavor of its own: Like hallucinogens for your taste buds, well-aged chèvres distort the tongue's sensitivities, affecting the flavor of all foods eaten afterward. Wine must be especially potent to pair with aged chèvre, and even the boldest brews can have their tastes twisted.

The Beauty of Aged Chèvre

Relatively unknown in North America, this class of cheeses includes some of France's most famous *fromages*: ash-coated and pyramid-shaped Valençay; Sainte Maure—pierced with a blade of straw (the industrial version of Sainte Maure features plastic straws!); and small, moldy Crottin are all aged chèvre cheeses. Perhaps the only well-known North American aged chèvre is Humboldt Fog, a creamy, ash-ripened goats' milk cheese from Humboldt County, California.

When aged chèvres ripen, they develop the most interesting rinds of all aged cheeses. Ripened with *Geotrichum candidum*, their rinds form wrinkles on wrinkles that many North American cheesemakers scorn as "toad skin," but that European connoisseurs consider to be intricate, beautiful, and a sign of a well-made cheese.

Also known as aged lactic cheeses, this style of cheese is made primarily with goats' milk; no other animal's milk responds so elegantly. The ripening effects of *Geotrichum* can cause lactic cows' milk cheeses to become too liquidy if subjected to the same process, so cheesemakers usually don't make these cheeses with it (Saint-Marcellin, recipe below, is a notable exception). Goats' milk has just the right composition and is a perfect match for this method: The nature of goats' milk responds well to the extended fermentation that results from the long, slow process.

Essentially chèvre that is formed, salted, and aged, these cheeses are also particularly easy to make at home. Because their methods are undemanding, relatively relaxed, and not too time-sensitive, these cheeses also fit well into the rhythms of home life. Considered to be the ideal of farmstead cheeses, aged chèvres are made at small farms all over France.

Encouraging White Rinds Naturally

Traditionally made with raw goats' milk, these cheeses grow beautiful white *Geotrichum candidum* rinds as a result of patient practices that encourage the growth of the indigenous fungal cultures of the milk. If the cheeses are rushed, the native *Geotrichum* will not get a head start, and other unwanted fungal and bacterial cultures will grow on the rinds and result in a less picturesque cheese.

Bloomy white rinds come naturally to these cheeses because the long, slow fermentation of the

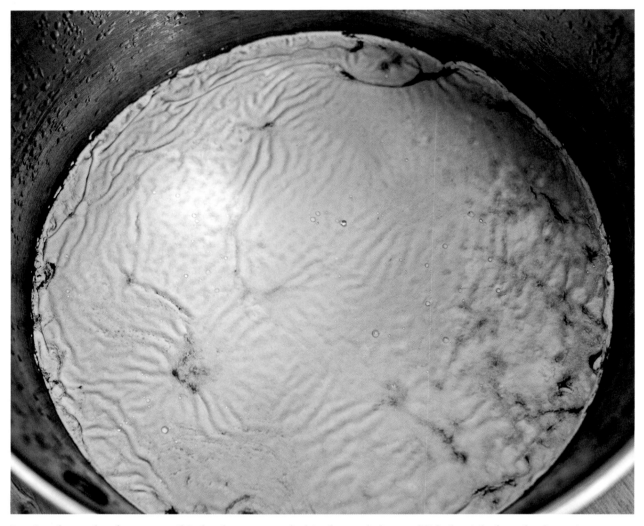

Leaving the curd to ferment until it develops a coat of white fungus helps establish the right fungal cultures in aged chèvre. This image shows the fungal development after 4 days of fermentation.

curd gives a boost to *Geotrichum* over other unwanted cultures: Left to ferment for over 2 days, the natural bloom of *Geotrichum* in raw milk will begin to spread over the surface of the curd in the pot.

Relying entirely on raw milk's indigenous *Geotrichum*, however, may give unpredictable results. Kefir, with more active *Geotrichum*, is a more reliable source of the fungus. Using kefir as a starter culture to make these cheeses will help the curd develop a thick white fungal coat quickly. Backslopping with whey from a previous batch of aged chevre will also ensure a healthy population of *Geotrichum*.

Freeze-dried DVI starters do not have the diverse ripening cultures that these cheeses need to develop their most unique appearances and flavors—contemporary cheesemakers who make aged chèvre with pasteurized milk and DVI starters must also add DVI *Geotrichum* or *Penicillium candidum* as ripening cultures.

A room-temperature curing early in the cheese's development further encourages the development of *Geotrichum*. This fungus, already active as a result of the extended fermentation, will cover the surfaces of cheeses curing at temperatures between 50 and

68°F (10 to 20°C) but not those ripened in cooler climes. Only once the right fungus has established itself on the cheeses are they moved to a cooler cave to age; the cooler conditions allow a more mellow ripening that helps develop the finest possible flavors in the cheese.

The humidity in the curing space (as well as the salt content of the cheese) affects the moisture content of cheese, which affects the development of wrinkles on the rinds of aged lactic cheeses. A drier curing space leads to the growth of finer, more intricate ripples on the rinds, while more humidity makes a more rounded and wavelike rind that more resembles a rhinoceros's skin. Do not let the curing space become too dry, however; a dried-out aged lactic cheese will lose its *Geotrichum* growth and come to be dominated by *Penicillium roqueforti*.

The Basic Method

Crottin, Valençay, Saint-Marcellin . . . all of these diverse aged cheeses start in a similar way. The basic process can be summed up as follows:

The milk is warmed, cultured, and lightly renneted, then left to ferment for a significant period of time at room temperature—usually 2 days, but often longer. The long, slow fermentation begins the breakdown of the milk's fats into flavorful fatty acids that give unique notes to these cheeses. The curd, made with a quarter dose of rennet compared with regular-curd cheeses, is soft and yogurtlike in texture and sinks to the bottom of the pot. The whey that rises over the sunken curd becomes covered with a slight film of *Geotrichum* fungus.

The curd is then strained from the pot, just like chèvre. But instead of draining in cheesecloth, the curds are ladled into forms and allowed to settle for a day to take the forms' shape. Once the curds have firmed, the cheeses are removed from their forms, salted and air-dried to control their moisture content, then left to cure at high humidity and room temperature for a week to grow their natural fungal coat. The cheeses are then placed in a cooler ripening environment to complete their aging; as they age, the fungal coat begins to soften their flesh from the surface toward the interior, resulting in a smooth and creamy paste (more on the effects of ripening fungi in chapter 16, White-Rinded Cheeses.

Unlike other cheeses, aged chèvres can be eaten at any stage of their development: fresh out of their forms and not even salted (this ultrafresh cheese is known as *faiselle*); with the fungal coat just starting to appear; with the rind well developed and paste turned molten; or even dried to wrinkled crumbs of their former selves. And at every stage of development they have strikingly different flavors.

The Many Hats of Aged Chèvre

Depending on the shape that they are given, aged chèvres can take on many distinct characters. The curd can also be pre-drained before forming, allowing cheesemakers to shape it into unique forms much like clay!

As well, different aged chèvres are given unique treatments, including leaf wrapping and ash ripening, that influence the ripening in interesting ways: Each unique method of handling results in a distinctly different cheese.

Shape

The shape of the chèvre greatly influences the way it ripens, and the different forms, be they logs or Camemberts, small pucks or pyramids, even ample breasts complete with nipples (a rare French cheese known as *Seine de Nounou*—"the wet-nurse's breast"), all ripen differently. In particular, the relationship of surface area to volume is important in the development of these small, surface-ripened cheeses: Cheeses with a higher surface-area-to-volume ratio will tend to ripen more quickly from the exterior than will those with a smaller ratio. Larger cheeses, with lower surface area compared with their volume, tend to ripen more from the interior, whereas smaller or flatter cheeses, those with square edges, or long and cylindrically shaped, tend to ripen fungally from the exterior.

A charcoal-ripened aged chèvre showing the effect of the fungal cultures melting the flesh of the cheese.

The many shapes of aged chèvre. These freshly salted cheeses were all made from 1 gallon of goats' milk.

Pre-Draining

Some aged chèvres are pre-drained before forming. Pre-draining gets the whey out of the curd, firming it up and giving it a greater workability. The malleable curd can be pressed, doughlike, into cookie-cutter forms, or hand-molded, claylike, into a vast array of shapes. To pre-drain curds, drain them for 1 to 2 days, salt them, then let them drain again for 1 or 2 days until they're firm and beginning to show sign of *Geotrichum* growth. The firm curd can then be pressed into its forms into its desired shape, cured for a week to develop its rind, then cave-aged like other aged chèvres. The recipe below for Mason Jar Marcellin involves a pre-draining stage before the curd is pressed into a Mason jar to age.

Ash Ripening

Ash, actually finely ground charcoal, can be sprinkled on the surfaces of aged chèvre cheeses to give them a bold black-and-white contrast. Typically, ash is added before the cheeses sprout their fungal coats; this way, the fungus grows through the ash, creating a crisp black line between the cream-colored fungal coat and the pure white goats' cheese. Valençay (recipe below) is a beautiful example of an ash-ripened aged chèvre.

Ash can also be added to the interior of aged chèvres. This is done when the cheeses are being formed: The forms are filled ⅔ of the way up with curds, the curd is allowed to drain and dry for several hours, fine-ground charcoal is sprinkled over

Natural wrapping papers including leaves and hornet's nests vastly improve the appearance and the flavor of aged chèvre.

top of them, and the forms are filled to the top with curds. This treatment leaves a clean black line through the interior of the cheese. A recipe for making ash is provided at the end of this chapter.

Leaf Wrapping

Many different types of leaves can be used to wrap chèvres. French cheesemakers often use true chestnut (not horse chestnut—it's mildly poisonous) and grape leaves. Maple leaves work, too, and you can use native plants from your forest to create distinctive bioregional cheeses.

Favorite wrapping leaves of mine are the cornhusks used to make tamales, which can be used to form chèvre into a corncob shape. I also appreciate wild-nettle-wrapped cheeses, as the nettles bring their own unique ecosystem and brilliant color to the rinds of cheeses made with them.

The leaf wrappings contribute to the ripening cheese in several ways. The leaves help keep the cheeses moist, encouraging them to ripen more evenly. The leaves add flavors to the cheeses from the tannins they contain. They give the cheese a leaf-impression surface texture, as well as beautiful color. And the leaves also create a handy package that protects the cheese from handling and pests. Leaf wrappings are usually applied to cheeses only once they have grown their fungal coats.

- RECIPE -
CROTTIN

Crottin (pronounced *CROW-ten*) is the easiest of aged chèvres to make: It is simply a little log of chèvre, aged until it's good and moldy.

The name is a tongue-in-cheek reference to this cheese's unusual appearance. French for "little turd," Crottin looks like a moldy horse turd plucked from a pasture! But don't be fooled; this cheese defies its scatological appearance and is definitely better sampled than trampled.

Geotrichum candidum is the fungus of choice for ripening Crottins. Because this is found in raw milk, traditional French cheesemakers didn't have to add any fungal culture to their cheeses; the raw milk *Geotrichum* bloomed on its own. If you're making Crottin with pasteurized milk, you must add *G. candidum* to ensure good rind development. Kefir is therefore an excellent starter culture to use as it contains exactly this fungus.

To make this cheese, you first warm goats' milk, then culture and rennet the warm milk and leave it to ferment and set a soft curd for 2 days, until a slight film of fungus grows on its surface. Ladle the curd into small cylindrical forms for 24 hours, then salt the cheeses and cure them at room temperature for 1 week to develop their natural *Geotrichum* coats. Once the *Geotrichum* is established, transfer the cheeses to a cooler cave to slow their development, encouraging more interesting flavors and textures.

INGREDIENTS

1 gallon (4 L) good goats' milk
¼ cup (60 mL) active kefir or active whey (if the milk is pasteurized, use kefir to assure the development of *Geotrichum*)
¼ dose rennet (I use ¹⁄₁₆ tablet WalcoRen calf rennet)
4 teaspoons (20 mL) good salt

EQUIPMENT

1-gallon (4-L) cheesemaking pot
Slotted spoon
5 short cylindrical goats' cheese forms
Draining table
Curing space at 50 to 68°F (10 to 20°C) and 85% humidity
Cheese cave at 40°F (4°C) and 90% humidity

TIME FRAME

2 weeks–2 months

YIELD

Makes 5 small Crottin cheeses

TECHNIQUE

Warm the goats' milk slowly to around 90°F (32°C)—the temperature of a baby bottle—then turn the heat off.

Add the kefir or whey to the pot along with the quarter dose of rennet dissolved in water. Gently stir the warm goats' milk to incorporate the rennet and culture.

Leave to ferment at room temperature until the *Geotrichum* blooms: Cover the pot, and leave it out, at 68°F (20°C), until the signs of fungal growth appear atop the curd, usually after 2 days of fermentation. During this time, the curd will acidify, firm up into a yogurtlike consistency, and sink under its whey to the bottom of the pot. A slight film of fungus will soon bloom over the curd—a sure sign that the *Geotrichum* fungus is established.

Ladle the curds into the forms: With the slotted spoon, transfer the curds from the pot to the Crottin forms atop the draining table. Fill the forms right to their brims.

Let the curds drain for 24 hours. Flip the young cheeses at some point, whenever convenient; the curds will be very soft, so handle them gently.

Salt the Crottins: Carefully take the Crottins out of their forms and salt their surfaces, adding 1 teaspoon (5 mL) salt to each cheese.

To make Crottin, ladle the curd into small cylindrical forms; flip the cheeses; then salt them before curing and aging.

Leave the salted cheeses to air-dry for 1 day at room temperature in a fly-free but not overly humid space. Flip the cheese once or twice as it dries to allow the moisture pulled out by the salt to flow away.

Cure the cheeses: Leave them for 7 to 10 days at 50 to 68°F (10 to 20°C) in a humid and fly-free draining space. Gently flip them daily to keep them from sticking to their draining mats. *Geotrichum* growth will begin to bloom from the cheeses' rinds.

Age in a cheese cave: Once the cheeses have an established fungal coat, transfer them to a cooler cheese cave. Flip them every other day as they age. Crottins can be eaten at any time from 2 weeks to 2 months of age. They are often removed from their caves and dried down once they have begun to liquefy, giving them a firm and chalky texture.

- RECIPE -

MASON JAR MARCELLIN

Saint-Marcellin is a rare aged lactic cheese made with cows' milk.

Cows' milk is generally not used for aged lactic cheeses for several reasons: First, it does not make as flavorful a cheese as goats' milk does (though the flavor of an aged lactic cow cheese is still comparable to Camembert); second, the cream from cows' milk rises to the top of the curd during the daylong fermentation and can leave buttery bits in the cheese (goats' milk does not separate its cream during this time); and, most important, the lower-fat curd that results tends to

soften up excessively from fungal effects as it ages, resulting in "slip" rinds that can deform the shape of the cheese, and cause it to melt into a puddle as it ages.

The aging of Saint-Marcellin in terra-cotta pots solves the problem of slipping rinds. The glazed pots protect the cheeses as they age, allowing the delicate curd to soften up with the aid of *Geotrichum* fungus into a molten lavalike texture without melting away. The pots are usually 4 inches across and 1½ inches deep (10 cm across and 3 cm deep). The cheeses are formed to fit inside with little to no wiggle room and cured for a week like Crottin before being placed in the pots to finish.

You can also make a Mason Jar Marcellin—a recipe of my own invention—by pre-draining and salting the curd, then pressing it inside widemouthed Mason jars to age. The Mason jars, lightly sealed with their lids and left in a cool place, also make excellent mini cheese caves!

To make a Mason Jar Marcellin, start by making a lactic cheese with good cows' or goats' milk, kefir as a starter, and a quarter dose of rennet. The milk is left to ferment and set for 2 days or more until it develops a film of *Geotrichum* from the indigenous cultures of raw milk or kefir. The yogurtlike curds are then ladled into cheesecloth, hung overnight, then salted and left to air-dry for 2 additional days. The pre-drained and salted curd is then packed neatly into Mason jars and left to cure at room temperature for 1 week to grow its natural *Geotrichum* coat. Once the cheeses are covered in fungal growth, put them in a refrigerator to age.

You can simply leave the cheeses in their pots to age. There is no need to flip them; they're meant to meld into the pot. And by regularly wiping down any moisture that accumulates on the Mason jar lids, you keep the aging environment at the perfect level of humidity. Left undisturbed, the cheeses' fungal coat grows extra-thick and luscious, and their

To make Mason Jar Marcellin, pre-drain and salt the curd; drain the curd for an additional day or two; then pack the dry curd into Mason jars to cure and age.

flesh begins to soften. The cheeses can be eaten after 1 month of aging, but it takes nearly 2 for the cheese to melt to its core (you can observe the fungal effects on the ripening curd through the glass); once it does, Mason Jar Marcellin can be eaten just like *fondue*!

INGREDIENTS

1 gallon (4 L) good milk, cows' or goats'
¼ cup (60 mL) kefir or active whey
¼ dose rennet (I use ¹⁄₁₆ tablet WalcoRen calf rennet for 1 gallon milk)
Good salt

EQUIPMENT

1-gallon (4-L) stainless-steel pot
Slotted spoon
Cheesecloth
Colander and large bowl
5 widemouthed 8-ounce (240-mL) Mason jars ("shorties") with lids
Curing space at 50 to 68°F (10 to 20°C) and 85% humidity
Refrigerator at 40°F (4°C) for aging

TIME FRAME

30 minutes to start; 1–2 months

YIELD

Makes 5 Mason Jar Marcellin cheeses

TECHNIQUE

Slowly warm the milk to 90°F (32°C), or baby-bottle-warm.

Add the active kefir or whey and mix in lightly.

Add the rennet: Dissolve a quarter dose of rennet in ¼ cup (60 mL) water, and gently mix it into the milk.

Ferment until *Geotrichum* blooms: Cover the pot and leave it out, at room temperature, for at least 2 days. During this time, the milk will sour, firm up into soft curd, and sink under its whey to the bottom of the pot.

Pre-drain the curds: With the slotted spoon, transfer the curds, from the pot to a cheesecloth-lined colander. Tie the cheesecloth up over the curds, and cover it with a clean towel to protect it from flies.

Let the curds drain for 24 hours in a colander over a stainless-steel bowl or hung from a spoon.

Salt the curd: Apply 1 tablespoon + 1 teaspoon (20 mL) of salt to the curd. Mix it roughly throughout.

Let the curd continue to drain and air-dry after salting for 2 more days.

Pack the cheeses into their Mason jars: Once the curd is drained and dry and crumbly in texture, pack it neatly with a spoon into the Mason jar pots. The cheeses should fill them about halfway to the top. Level the top of each cheese to give it a flat surface. And cover the Mason jars lightly with their lids but do not seal.

Cure the Marcellins: Leave the pots of cheese to cure at 50 to 68°F (10 to 20°C) for 7 to 10 days to develop their fungal coats. The *Geotrichum candidum* fungus from the raw milk or kefir will bloom on the surfaces of the cheeses. Open the pots daily to check on the cheeses' progress, to let them breathe, and to wipe off any excess moisture that accumulates on the lids.

Put the jars in a cool place to age: Once the cheeses are covered in wrinkled fungus, place the pots into a refrigerator to age. Because the cheeses are aged in sealed Mason jars, there is no need to control the humidity of the cheese cave environment.

Check up on the cheeses twice weekly to be sure that their aging conditions are just right. Open up each jar, and wipe off any moisture from the lid every time you check on them. Mason Jar Marcellin cheeses will be ready in 4 to 8 weeks.

- RECIPE -
VALENÇAY

This pyramid-shaped cheese is made in a nearly identical manner to Crottin. But the shape this cheese is given defines its identity as a uniquely delicious aged chèvre.

Call it pyramid power if you'd like, but the real reason Valençay develops superlative flavors is because it is never flipped! The cheese always rests on its bottom end and as a result develops two completely different ecologies that combine to form one exceptional cheese. The exposed parts develop a *Geotrichum* fungal coat, while the bottom, never getting a chance to breathe, develops a *Brevibacterium linens*–influenced washed rind (see chapter 18).

Both the *Geotrichum* and the *B. linens* cultures can develop naturally on their own if raw milk is used to make this cheese. If you begin with pasteurized milk, using kefir as a starter will help establish both of these integral cultures. If you use pasteurized milk and rely on packaged cultures to make Valençay, you will have to add DVI *Geotrichum* and *B. linens* cultures to the milk in addition to the mesophilic starter culture—that's three different packaged cultures (and many cheesemaking guidebooks also suggest the use of DVI *Penicillium candidum* and yeasts). And yet many of the yeasts and bacterial cultures that this cheese needs to age will likely still be lacking!

This cheese can be ashed, or left to ripen on its own. If it is ashed, however, the corners of the pyramid, with less fungal growth because they are drier, will show their black color in beautiful contrast with the white fungal growth: Valençay is one of the most striking of cheeses.

INGREDIENTS

1 gallon (4 L) good goats' milk
¼ cup (60 mL) active kefir or active whey
¼ dose rennet (I use ¹⁄₁₆ tablet WalcoRen calf rennet)
Salt
Cheese ash (see recipe below)

EQUIPMENT

1 gallon (4-L) cheesemaking pot
Slotted spoon
5 pyramid-shaped Valençay forms
Draining table
Small fine sieve, or wire mesh tea ball (my favorite for sifting ash)

Curing space at 50 to 68°F (10 to 20°C) and 85% humidity
Cheese cave at 40°F (4°C) and 90% humidity

TIME FRAME

2 weeks–2 months

YIELD

Makes 5 Valençay cheeses

TECHNIQUE

Slowly warm the goats' milk to 90°F (32°C), or a baby-bottle-warm temperature, then turn the heat off.

Add the kefir or whey along with the quarter dose of rennet dissolved in water. Gently stir the warm goats' milk to incorporate the rennet and culture.

Leave to ferment at room temperature until *Geotrichum* blooms: Cover the pot and leave it out, at room temperature, for at least 2 days, until a slight film of fungus grows atop the curds. During this time, the curd will acidify, firm up into a yogurty consistency, and sink under its whey to the bottom of the pot.

Ladle the curds into the forms: With the slotted spoon, transfer curds from the pot to the Valençay forms atop the draining table. Fill the forms right to their brims. Balance the top-heavy Valençay forms atop cylindrical forms if need be to keep them steady.

Drain 24 hours. Simply leave the cheeses to drain, protected from flies; there is no need to flip the curds as they take their pyramid shape.

Salt the cheeses: Take the cheeses out of their forms, and apply 1 teaspoon (5 mL) of salt to the surfaces of each. Spread the salt evenly on the bottom and sides of the pyramids but not the top.

Air-dry the cheeses at room temperature for 1 day. Use a draining rack in a fly-free draining space.

Sift the ash over the cheeses: Carefully pick up each cheese, and sift the fine-ground charcoal over every surface but the bottom. Scoop out some of the charcoal into a fine sieve, then sift it

To make Valençay, ladle the curd into pyramid shapes to drain; salt the cheeses that result; then coat them with charcoal.

through by lightly tapping the sieve with a finger. Use only enough to create an even black coat; too much charcoal can give the rinds a grainy texture.

Cure the cheeses for 1 week: Leave the ashed cheeses in a humid curing space at 50 to 68°F (10 to 20°C) for a week to allow the *Geotrichum* to bloom on the surfaces of the pyramids. Do not flip the cheeses, but gently lift them so that they do not stick to their mats. The cheeses are ready to be placed in their cave when their rinds are covered with white fungus and have just begun to wrinkle.

Place the cheeses in a cheese cave to age. Don't flip them, but do lift them up every other day to be sure that they don't stick to their aging surfaces. Valençays can be eaten at any time from 2 weeks to 2 months of age.

– RECIPE –
CHEESE "ASH"

The ash you often see gracing the surfaces of cheeses is not ash at all, but finely ground charcoal!

Some cheeses such as Camembert were traditionally ripened in ash, which, in influencing the acidity balance of cheeses, can cause them to ripen more slowly. Ash was also used in Morbier: Here the cheese made from the morning's milking would sit around waiting for the cheese made from the evening's milking to be added on top; cheesemakers would sprinkle ash on the sitting curd to prevent flies from being drawn to it!

Today, however, these two cheeses are made with charcoal and not ash. Charcoal, unlike ash, is inert; its effect on cheeses is limited to the aesthetic. Used primarily to provide color contrast in white bloomy-rinded cheeses, charcoal has no significant impact on a ripening cheese.

The process of making your own "ash" is the process of making charcoal.

You can use many different types of wood to make it: the traditional French cheese ash is made from a certain species of pine tree, but any fruitwood or hardwood would work.

Burning the wood in oxygen-starved conditions transforms it into charcoal. On the island where I farm, Japanese Canadian settlers once tended massive earthen kilns in which they cooked logs into charcoal, which they sold to nearby salmon canneries before they were interned because of public suspicion, racism, and greed during the Second World War. Their charcoal-making process involved building a massive fire inside a pit, then covering the pit with earth, starving the fire of air, and allowing the wood to slowly transform, over the course of several days, into charcoal.

Grinding homemade charcoal yields cheese ash.

You can easily replicate the process at home on a small scale in a woodstove or fire pit. Though you cannot create the oxygen-starved conditions that wood needs to turn to charcoal just by controlling the dampers of the woodstove, you can wrap wood sticks well in aluminum foil and place them in the glowing coals. Subjected to high temperatures within the fire, but limited in their exposure to air, the wood will change into charcoal in a very short time. Once cooled, the charcoal is finely ground and put through a sieve to remove any coarse bits. You can then give your cheeses a surface coating of "ash" by sifting the fine charcoal onto the rinds of your cheeses.

When sourcing wood for transforming into charcoal, be sure to use well-seasoned branches of hardwoods or pines, or fruit tree prunings. And find branches consistent in size to ensure an even charring.

INGREDIENTS
Good wood about ½ inch–1 inch (1 cm–2.5 cm) in diameter and 10 inches (25 cm) long

EQUIPMENT
Woodstove or fire pit
Aluminum foil
Mortar and pestle
Sieve

TIME FRAME
30 minutes

TECHNIQUE
Start a fire. Of course this is something you can do at a bonfire, while camping, or in a woodstove at home.

Wrap the wood well in aluminum foil, being careful to cover all sides. Any exposed wood will turn to real ash.

Place the wrapped wood into the fire, on a bed of coals. Leave them for only 5 to 10 minutes. Carefully remove the charred wood from the fire, and let the pieces cool for several minutes before unwrapping them.

Grind the charcoal extra-fine: Break each stick into inch-long (2.5 cm) pieces to make grinding easier but also to ensure that each piece of wood is well charred. In a mortar and pestle, grind the charcoal well. The final consistency will be fine like icing sugar; this takes only a few minutes. If there are any larger-sized particles, they will be felt on the teeth.

Sieve the charcoal: Pass the charcoal through a fine sieve to remove any big bits of charcoal. Put these larger bits back in the grinder to grind smaller.

Store the charcoal: Carefully transfer the charcoal into a container. Sealed shut, it will last for years.

CHAPTER THIRTEEN

Basic Rennet Curd

The word *curd*, like *dough*, refers to a work in progress, an unfinished cheese still being molded by the hands of the cheesemaker. Curd is the gel that forms when rennet is added to warm and sour milk. Curds are cut from the mass of curd, then stirred to rid them of their whey. Only when they are strained from the whey into their forms to take their final shape do the curds become cheese.

The basic rennet cheese that evolves from the curds tastes simply of milk, but don't let their simplicity fool you: The curds have the potential to become a dizzying array of complex-tasting aged cheeses. Like a plain canvas, curds are infinitely impressionable, and every different way that they are handled and aged results in a distinctly different cheese.

Basic Methods

The basic method for making curd is the foundational technique from which nearly all the remaining cheeses in this book evolve. Learn this method well and you'll be on your way to making mozzarella, blue cheese,

A fresh rennet cheese made from basic rennet curds.

Camembert, and feta; even firmer cheeses such as Gouda and cheddar start as basic rennet curds.

Once you're familiar with the technique, making an aged cheese will be a breeze. The following is a summary of how a basic rennet cheese (and the basic rennet curds that make a cheese) can be made into different aged cheeses.

Fermenting and Stretching into Pasta Filata Cheeses

An unsalted rennet cheese can be left to ferment in its whey to develop its acidity to the point where, when submerged in hot water, its curd stretches. Pasta filata, or stretched-curd, cheeses such as mozzarella are made by fermenting the curd, then stretching it in hot water into various shapes and textures. More information on pasta filata cheeses can be found in chapter 14.

Salt-Brining

A rennet cheese can also be submerged in a salty brine to age; ripening a rennet cheese in brine will transform it into feta (see chapter 15).

Encouraging White Fungus

The indigenous fungal cultures of the milk can be encouraged to help the cheese to develop a white rind. Cheesemakers can create conditions under which milk's native *Geotrichum candidum* fungus thrives, resulting in the growth of fungal coats that give Camembert and other cheeses white rinds. More information on making white-rinded cheeses can be found in chapter 16.

Encouraging Blue Fungus

Cheesemakers add spores of *Penicillium roqueforti* to the curd to help it turn blue. When handled in the right way, *P. roqueforti* gives rennet cheese beautiful blue rinds or striking blue veins. More information on blue cheese is provided in chapter 17.

Washing Rinds

A rennet cheese can be washed with salty whey as it ages to keep fungus in check; the regular washing encourages the development of a microbial ecology that will transform the cheese into a stinky washed-rind cheese. Chapter 18 explores the process of making washed-rind cheeses.

Alpine Cheeses

In a variation on the basic rennet curd method, the curd of Alpine cheeses is cut to a very small size to give off more whey, then cooked and pressed into very firm and large cheeses. More information on cooked cheeses is provided in chapter 19.

Washing the Curds

To make Gouda, basic rennet curds are washed in hot water to firm them up. The firmer curds are pressed into rounds of cheese and aged, usually covered in wax. Chapter 20 covers washed-curd cheeses.

Cheddaring

Cheddaring is a third method of making rennet cheeses firm. To make a cheddar cheese, basic rennet curds are given a special treatment known as cheddaring; this method of handling involves letting rennet curds knit together into a loaf, then slicing the loaf and stacking and restacking the slices of curd to press them firm. The firmed curds are then milled, salted, and pressed into a cheddar cheese. The cheddaring process is explored in chapter 21.

Transforming Milk into Curd

The process of making a basic rennet cheese is a transformation in three stages: Milk is first transformed into curd; the curd is then cut and firmed into curds; and finally the curds are formed into a cheese. For each stage of the transformation there is a series of steps to be followed. The next three sections explore how and why cheesemakers take these steps.

The first stage of the transformation invokes bacterial cultures, rennet, and warmth to turn milk into curd: When souring milk comes in contact with the rennet enzyme in warm conditions, it sets into a

A basic rennet cheese is regularly washed with whey to help it age into a washed-rind cheese.

jelly-like curd. The following are considerations to keep in mind with respect to the milk, the culture, the rennet, and the warmth to get a good set.

Sourcing Milk

Raw milk from well-raised, pastured animals will respond best to this process. Good milk that has been pasteurized will respond nearly as well, but your average grocery-store-bought, pasteurized, homogenized milk from confined cows will fail.

Aim to use milk that's as fresh as possible. It is best to use milk on the same day it is milked, but under refrigerated conditions it will keep its best cheesemaking qualities for up to a week. After that the milk can play host to unwanted microorganisms that can cause problems in a cheese.

This recipe works with cows' milk, goats' milk, sheep's milk, or buffalo milk, but each will result in a slightly different cheese.

The timing of the recipe is accurate for using fresh raw milk from pastured animals. If you use more processed milk, or milk from animals that are fed only haylage, silage, or grain, you may find that it takes longer to achieve clean break or for the curds to develop their desired firmness. However, if your milk is older and therefore more acidic, the process will take less time.

Warming the Milk

Cheese happens in the warm conditions inside calf, kid, and lamb stomachs. When making cheese in a pot, cheesemakers mimic the natural conditions in which this process is most effective by keeping the cheesemaking pot at around body temperature.

The basic rennet curd process evolves best at temperatures around 90°F (32°C). This is the temperature at which the starter cultures are most active, and at which the natural rennet enzyme most effectively sets milk. If the temperature of the process is too low, the milk will acidify more slowly, and the rennet will take longer to set; at higher temperatures, the bacterial cultures will develop acidity more slowly, but the rennet will be more sensitive to the acid, and the curd will set unpredictably. Once

the milk is warm, the pot is kept at around 90°F for the length of the make, until the curds are pitched and ready to form.

Adding Culture

Culture breathes life into milk and helps it evolve into cheese. Culture also develops the acidity that milk needs to develop into curd, protects cheeses from unwanted microbial growth, and helps them develop particular flavors as they age.

Culture is added in several different ways: with whey drained from yogurt or saved from a previous cheesemaking session; with active kefir; or with freeze-dried DVIs. Regardless of which type of culture you use, milk will evolve into cheese in a similar manner; it is only as cheeses age that the differences between the cultures begin to become apparent.

Traditionally, whey from a previous batch of cheese is saved and reused as a mother culture for the next batch of cheesemaking. In saving whey from one batch of cheese for the next, cheesemakers evolve and maintain the appropriate community of cultures for the conditions under which their cheese is made. It is only possible to practice this method if you're using raw milk, which contains a diverse collection of native cultures that nourish and maintain the mother culture. I find that you don't need to keep separate whey cultures for making different types of cheese: so long as there is a diversity of microorganisms in the whey, the cultures can adapt to the making of any style of cheese. More information on keeping a whey starter is provided in appendix B.

Kefir can also be used for starting rennet cheeses. It contains a community of microorganisms well suited to cheesemaking. It can help pasteurized milk establish a bacterial profile not unlike good raw milk and can also help raw milk to grow a strong microbial community that helps to establish appropriate ripening regimes. Kefir will remain active for up to a week in the refrigerator, and can be kept until you are ready to make cheese; however, I find that it is best to regularly feed the kefir culture, and to give the culture fresh milk the day before you make cheese to ensure that the culture is in its best shape for cheesemaking. Active kefir can simply be added straight to the milk as a starter.

Whey strained from commercial yogurt can be saved and used for making a rennet cheese, but it is best for unaged cheeses. Whey from commercial yogurt contains several different strains of laboratory-raised mesophilic and thermophilic cultures that make it a suitable starter culture for many different styles of fresh cheese; it does not, however, contain the broad spectrum of cultures that are needed to help cheeses age well. For example, whey strained from commercially prepared yogurts will not contain the *Geotrichum candidum* culture necessary for the development of white rinds.

Contemporary cheesemakers use freeze-dried DVIs to culture their cheeses. They measure out the prescribed dose, sprinkle it atop the pot, leave it there to hydrate for several minutes, then mix it into the warm milk. Depending on what style of cheese is being made, any one of dozens of different DVIs can be used as a starter. If you wish to culture your cheeses with DVIs, I don't have much advice to offer. Every other cheesemaking guidebook offers guidance on the subject; I, however, refuse to recommend their use.

Incubating the Milk

After culturing, leave the pot in a warm place to incubate for 1 hour. The simplest approach to incubating the souring milk is to leave it on top of the stove where it was warmed: The residual heat will help keep the milk warm and encourage its added cultures to thrive. Moving the pot of cheese to a separate incubation chamber, as many bakers do with rising breads, is not recommended for cheesemaking; to encourage curd development, it's best to leave the pot where it is and disturb the developing curd as little as possible.

The temperature of the pot of milk can be kept even more stable during the incubation periods by placing a lid on the pot and wrapping it in towels. If the temperature of the cheesemaking pot drops, you can turn on the stove for several seconds underneath the pot; the added heat will help keep the stove a warm place to incubate your cheese. It is acceptable for the pot temperature to drop several degrees as it incubates; too much of a decrease in temperature, however, can cause the development of the cheese to slow.

If you are making cheese with just a gallon of milk, you may find that you will have difficulty keeping it warm. A larger volume of milk holds on to its heat longer than a smaller one; making cheese within a warm-water bath can help keep a smaller batch of cheese warmer for longer.

Renneting

After this first hour-long incubation period the milk is renneted. Rennet is added to milk by first dissolving and diluting it in clean, unchlorinated water, then pouring the rennet water over the milk. There is no need to be concerned about diluting the milk with the water; the added water will flow away with the whey and won't get in the way of the cheese.

Powdered rennet is weighed or measured with a measuring spoon. Liquid rennet is added drop by drop. Tableted rennet is cut in segments. Each different brand of rennet calls for a different dosage for basic rennet curds, so consult the instructions on the package before using your particular rennet.

I use WalcoRen brand calf rennet tablets, available at most cheesemaking supply houses, and recipes are written out with exact dosage for this specific rennet as an example; other types of calf rennet can be used accordingly to their standard of use. If you use microbial rennet, the dosage will be different, the coagulation time will be different, and the curd will firm at a different rate. The cheese that results will also be discernible in taste and texture from one made with calf rennet. Cheesemakers beware: If you choose to use a calf rennet substitute, these recipes might not work as written.

Following renneting, another hour's incubation at 90°F (32°C) is then needed to allow the milk to set into curd. Be sure that the milk is disturbed as little as possible during this second incubation to ensure that the developing curd isn't damaged.

Transforming Curd into Curds

Once the milk has set, the curd is cut into smaller curds to encourage it to give off its whey; the curds are stirred to keep them from knitting together until they reach their desired firmness. But before all that is done, the cheesemaker must judge the readiness of the curd with the clean break test.

The Clean Break Test

With your eye, you can observe the gelling, sinking, and shrinking of the curd to determine its readiness. The curd is ready when you see a slight bit of whey around the edge of the pot, showing that the curd has begun to shrink. And the curd will begin to sink and will show a slight puddling of clear yellowish whey upon its surface.

The clean break test involves sticking your finger into the curd at a 45-degree angle. You should feel a pop as your finger breaks the surface tension of the curd—it's a rather satisfying sensation, as it tells you the curd is ready. Then lift your finger straight up. The curd should rise above your rising finger, then cleave cleanly in two.

If the curd has not set after an hour, wait another 15 minutes, keeping the pot warm, and try again. If after 2 hours the curd has not set, consider what you may have neglected in the previous stage with respect to the milk, the culture, the rennet, and the warmth. If the milk is good, the culture is active, the rennet is fresh, and the pot is kept warm, the curd will set in an expected time frame—you can almost set your watch by it. But if one of these ingredients is not as it should be, then the curd may not set.

Cutting the Curd

The curd is cut into ¾-inch (2 cm) curds to encourage the release of whey. The whey flows from all the extra surface area given to the curds by the cutting, and this results in a firmer curd.

The curds are cut to size in three series of slicings. First the curds are cut into ¾-inch (2 cm) slices by pulling the blade of the knife from one side of the pot to the other; the slices of curd are then cut into columns by making another series of slices at right angles to the first. A final series of horizontal slices cuts the curds into ¾-inch (2 cm) cubes.

When cutting curd in a round pot with a square knife, a horizontal cut is challenging to achieve. But that's okay; the curds don't care if they're square, and a series of cuts on a shallow angle from both sides of the pot can give the curd the extra surface area it needs.

Commercial cheesemakers make rennet curds in square vats that facilitate the horizontal cutting. And cheese harps, which consist of wires on a rack, are pulled through the curd to cut it. Cheesemakers have two sets of curd knives called cheese harps: one with horizontal wires and one with vertical. The vertical wires are used to cut the curd vertically in two directions perpendicular to each other, and the horizontal knives are used to cut the curds on the final plane.

Be sure that the curds have been cut consistently throughout the pot. If, while stirring, you see larger curds, cut them to size. Large curds give off their whey more slowly than small curds, and inconsistent curd size may lead to inconsistent moisture content and inconsistent ripening in a cheese.

Stirring the Curds

Once the curds have been cut, they are stirred to release their whey. Stir in a figure-8 pattern, changing the orientation each time so that no curd is left unstirred.

Stir the curds gently and slowly, with a wooden spoon or paddle, so that they do not break. A stainless spoon or a brisk stirring can break the curds, giving them more surface area and changing their texture.

A thorough stirring every 5 minutes will suffice to encourage the curds to firm. If, however, you leave the curds unstirred too long, they can settle to the bottom of the pot and knit together into a cheese. If this happens, don't fret; you can always break the curd up again with your hand and continue stirring, more frequently than before.

Judging the Curds

The curds are stirred until they have the firmness of a poached egg, which takes between 30 and 60 minutes, depending on the quality of the milk. They will also shrink considerably from their original size, and their cut edges will tighten and become more rounded.

You can judge the curds by pressing them between your thumb and forefinger: If they offer resistance and don't just break into pieces when you press them, they are ready for the next stage of transformation. Many cheesemakers compare the consistency of finished curd to a poached egg. When the curds have firmed, they are ready for the final transformation into cheese.

Transforming the Curds into Cheese

The final stage of transformation takes the curds and forms them into small rounds of cheese ready to eat or age. The firmed curds are pitched, then wheyed off, and finally placed into their forms. The curds are drained to allow them to knit together into cheese, the cheeses are flipped to ensure they have an even shape, and the formed cheeses are salted and air-dried.

Pitching the Curds

Once the curds have firmed, cheesemakers leave them to settle in the pot for several minutes. The additional time in the warm whey helps the curds firm up a bit more, and the settling of the curds to the bottom of the pot makes wheying off a lot easier. Letting the curds settle in the pot is known in some cheesemaking circles as pitching.

Wheying Off

With the curds at the bottom of the pot, the whey is ladled off, giving better access to the curds. You can even simply pour the whey off the pot into another vessel—you shouldn't even need a strainer to catch any curds that might flow with the whey,

as the curds are denser than the whey and will remain in the pot.

When wheying off is complete, the curds will be clearly visible at the bottom of the pot. Reserve the whey—it has many different uses explored in detail in chapter 22.

Forming the Curds

Cheese forms give the cheese its shape. Depending on the cheese you wish to make, you will want to have the appropriate-sized forms on hand. The curd can even be strained into cheesecloth and given a free-form shape.

The curds are strained into their forms, by hand. Fill the forms right up to their brims: The curds will give off whey as they form and will shrink into a cheese about a third of its original size. A half-filled form will result in a half-height cheese.

Soft and warm rennet curds knit themselves together without any added pressure. Only firmer-curd cheeses such as Alpine cheeses, cheddars, and Goudas need added weight to press their curds together.

Draining the Curds

The curds are left to settle into a cheese in their forms overnight. During this time, the forming cheeses are placed to drain at room temperature on a draining table that allows their whey to flow. It is important to protect the cheeses from flies as they drain: laying a cloth over the cheeses can keep flies at bay.

Flipping the Cheese

As the curds drain and coalesce in their forms into a cheese, they are flipped once or twice. After an hour or two of draining, the soft cheeses are carefully removed from their forms, flipped upside down, then placed back into their forms. Flipping ensures that the cheeses get an even shape top and bottom.

Salting the Cheese

Salt is applied to the cheeses at the end of the cheesemaking process. Once as much whey has been removed from the curd as possible, through cutting the curd, stirring the curd, straining the curd, and draining the curd overnight, added salt wicks away the remaining whey.

Salt is applied to the curd either by surface-salting or by brining. For home cheesemaking, I recommend surface-salting. Brining is helpful for pulling salt out of massive cheeses, but smaller cheeses made at home can be quickly and effectively salted by simply applying salt to their surfaces.

Drying the Curd

After salting, the cheeses are left to dry so that the salt can pull out all of the excess moisture. The cheeses are placed back on a lined draining table to dry. They are left out for about 24 hours at room temperature, and flipped once or twice to dry them evenly. They are ready when they are no longer wet to the touch.

If the edges of the cheeses have discolored, they have dried too long. Cover the drying cheeses with a cloth, if need be, to protect them from flies, and from becoming overly dried.

- RECIPE -
BASIC RENNET CHEESE

And now the recipe. The technique described herein is a distillation of the above considerations into a concise and, I hope, easy-to-follow method. In the cheeses that remain in this book that evolve from these curds, the recipes describe essentially the same steps, with added details for the steps that define each particular cheese. If there is any ambiguity in the recipes that follow, refer back to this recipe for clarity.

The cheese that results from this basic recipe is a semi-firm fresh cheese with a certain sweetness and distinctly milky flavor. It is, in essence, a concentration of milk, a coagulation of the most nutritious and delicious parts of the milk into a semi-solid form.

This cheese can be eaten as is, right out of the form. Milky, slightly salty, and with a chewy-squeaky texture, this fresh basic cheese makes for excellent snacking and cooking. Like yogurt cheese, this fresh cheese is eaten in vast quantities in Europe, the Middle East, and Central and South America. Outside certain communities, though, it is not very well known in North America.

INGREDIENTS

1 gallon (4 L) good milk
¼ cup (60 mL) kefir or active whey
Regular dose rennet (I use ¼ tablet WalcoRen calf rennet for 1 gallon milk)
Good salt

EQUIPMENT

1-gallon (4-L) pot
Wooden spoon
Long-bladed knife
Large bucket for whey
3 Camembert-sized cheese forms about 4 inches (10 cm) across and 3 inches (8 cm) deep
Draining rack set up with draining mat

TIME FRAME

4 hours focused cheesemaking; 3 days total involvement

YIELD

Makes 3 small rounds—about 1½ pounds (700 g) of cheese

TECHNIQUE

Warm the milk slowly to around 90°F (32°C). Stir the milk occasionally so that it does not scorch on the bottom, and check the temperature of the pot occasionally as it warms. At 90°F the milk will feel just barely warm to the touch.
Pour the active kefir or whey into the warm milk, and mix it in thoroughly with a spoon.

Incubate 1 hour. Keep the temperature of the pot maintained at around 90°F by covering the pot, wrapping it in towels, and leaving it in a warm place.

Add the rennet by dissolving the appropriate dose (according to the instructions on the package) in ¼ cup (60 mL) cold, clean, and dechlorinated water, then pouring the mixture over the milk. Mix in the rennet with a few seconds of gentle but thorough stirring.

Incubate 1 hour: Leave the renneted milk to rest for an hour to encourage the curd to set. Wrap

To make the curd, add culture to milk to acidify it; add rennet to set it; and judge its readiness with the clean break test.

the pot in towels to preserve its warmth, and check on it periodically to be sure it hasn't cooled. Warm the pot if necessary by turning on the stove for a moment, but do not disturb the developing curd by stirring the milk.

Check for clean break: About 1 hour after rennet-ing, the milk will set into curd. Perform the clean break test, by sticking your finger into the curd at a 45-degree angle, and lifting it straight up, to determine if the curd is ready to be cut.

Cut the curd: Once the milk has achieved a clean break, cut the curd to a ¾-inch (2-cm) size in a series of three cuts to increase its surface area. One series of cuts is made vertically. A second series is made vertically at right angles to the first series. And a third series is cut on an angle as close to horizontal as possible from both sides of the pot. Wait several minutes between each series of cuts to let the curds heal.

Stir the curds every few minutes for 30 minutes to an hour, to encourage them to expel more whey. As you stir, check the temperature of the milk by placing your hand on the side of the pot. If the curds have cooled, warm the pot for a moment on the stove as you stir, then turn the heat off.

Check for firmness: You want to stir the curds occasionally until they have the firmness of a lightly poached egg. Check their firmness

Cut the curd into ¾-inch (2-cm) curds, and stir the curds for 30 minutes, until they have the firmness of a poached egg.

periodically by scooping some out with your hand and pressing them with your fingers. When they offer some resistance to your touch, but are still silky and soft inside, they are ready to pitch.

Pitch the curds: Leave the curds to settle at the bottom of the pot for 5 minutes. Do not stir the pot during this time; the curds will naturally sink under the whey.

Whey off: Once the curds have settled to the bottom of the pot, the whey atop them can be poured off. Pick up the pot, and slowly pour the whey into another pot or clean bucket. Reserve the whey—it has many uses, explored in chapter 22, Whey Cheeses.

Form the curds: The curds can now be placed into their forms. Fill the 3 forms up with curds, by hand, such that each one is filled to the same height.

Drain the curds: Place the filled forms atop a draining rack to drain. And leave them out, at room temperature, covered with a cloth to keep flies at bay. As the curds sit in their forms, they will knit together into cheese.

Flip the cheeses: As soon as they are firm enough to handle, usually after 1 to 2 hours of draining, the cheeses are flipped so that they form evenly on both sides. After 24 hours of draining they will be fully firmed and can be taken out of their forms for salting.

Whey the curds off; strain them into their forms (and flip them once they've firmed); and salt the cheese that results.

Salt the cheese: Gently rub 1 teaspoon (5 mL) of salt over the surfaces of each cheese so that all sides are covered, and so that moisture is pulled from the cheese evenly from all sides.

Air-dry the cheese: The salted cheeses are placed back on the mat on the draining rack and allowed to dry for 24 hours. The salt attracts moisture from within the cheeses, and their surfaces will begin to glisten with small beads of whey pulled from within. Leave the cheeses out at room temperature overnight to allow that moisture to drain, and the cheeses to dry.

Flip the cheeses, when convenient, to allow any salty whey that has pooled atop them to drain. Once the cheeses have no more visible moisture on their surfaces, they are ready.

Keep the cheeses, or age them! The fresh cheeses can be kept, refrigerated, for up to 1 week or aged into mozzarella, Camembert, blue cheese, or a washed-rind cheese.

CHAPTER FOURTEEN

Pasta Filata Cheeses

Pasta filata, or "stretched-dough," cheeses are a class of rennet cheeses prepared in a distinct way that encourages acid development to the precise level at which, when submerged in hot water, the curd spins off fine cheesy fibers. The almost melted curd is stretched and shaped, cooled in brine, and generally eaten fresh, though there are pasta filata cheeses that are aged as well.

Stretching the Curd Around the World

Pasta filata cheeses achieve near-perfect plasticity—the ability to be stretched into a new shape and keep their form. The curd can be spun so finely that a single large cheese could very well stretch all the way around the world! And indeed these pasta filata cheeses have traveled to the ends of the world: Many traditional cheeses of the Middle East, Europe, Asia, and the Americas are made in this style . . . and, increasingly, North America is catching the thread. Traditional mozzarella, a celebrated pasta filata cheese, is now gaining in popularity (though American mozzarella—also a pasta filata—is still winning), and local versions can be found in grocery stores across America and around the world. Is there a town around without a Neapolitan pizza joint featuring true mozzarella?

Mozzarella is the most famous member of a family of cheeses that includes many distinguished relatives. Other Italian cheeses made in this style include Bocconcini, cream-filled Burrata, and long-aged, pear-shaped Caciocavallo. Arab cheesemakers, who may very well have invented this style of cheesemaking (they did, after all, bring the water buffalo to Italy!), make a lesser-known pasta filata cheese called Majdouli, a beautifully braided cheese peppered with black nigella seeds that looks like it belongs among the skeins of yarn in a wool shop. Mexico's favorite cheese, Oaxacan string cheese, is pulled apart into thin fibers and added to many different dishes. And I cannot forget to include processed string cheese, a Westernized, plastic-packaged, kid-friendly (but not so environmentally friendly) version of pasta filata cheese.

String Cheese Theory

It is acidity, or more precisely the particular way that the casein protein and calcium interact at a certain temperature and acidity, that gives pasta filata cheeses their magical ability to stretch.

At a precise acidity (between pH 5.2 and 5.4) and temperature (above 110°F—that's 43°C) the calcium in casein-based curd is released from the cheese. Calcium, which also gives strength to our bones, helps to keep cheeses firm. Without it, the casein protein loses its shape and realigns in a way that gives the curd unparalleled plasticity and workability. Once cooled, these cheeses lose their stretch and keep their sculpted forms.

Oaxacan string cheese shows the strongest development of pasta filata cheese's fibers.

This plastic curd is incredibly workable. It can be stretched and folded into nearly any shape. So long as the curd is kept hot, it can be stretched to an amazing length without breaking. The more it is worked, however, the tougher the cheese gets; the most tender of pasta filata cheeses, mozzarella, is shaped lightly and quickly to preserve its supple texture.

Cream does not contribute plasticity to pasta filata cheeses—only the casein protein gives these cheeses their stretch. Cream, however, plays an important role in giving softness to the strong curd. Pasta filata cheeses made with full-fat milk are considerably more luscious than those made with skimmed milk, which tend to have a more rubbery consistency. The more fat you pack into mozzarella, the better it gets, and that explains precisely why water buffalo milk makes the most sought-after mozzarella; with 6 percent fat content, buffalo milk makes the creamiest mozzarella.

Different Approaches to Achieving Acidity

The target acidity for stretching pasta filata cheeses can be achieved through one of two means. Cheeses can be either slowly fermented with the aid of bacterial cultures, or quickly acidified through the direct addition of a precise amount of acid, such as citric acid, lemon juice, or vinegar. These two different methods can be described, respectively, as slow and fast pasta filata cheeses.

Slow pasta filata cheeses are made acidic through the action of bacterial cultures that slowly ferment milk's lactose sugars into lactic acid. In this more traditional method, a cheese is made from milk according to the standard rennet method described in chapter 13. The curd is strained from the pot, then formed into a cheese, but instead of being salted to draw out its whey, the cheese is left to ferment in its whey and develops the necessary acidity

— 160 —

The ingredients for a fast and natural mozzarella: milk, lemons, and rennet.

Majdouli, a Middle Eastern braided string cheese, being salted in brine.

to stretch when heated. As a result of the lengthy bacterial fermentation, slow pasta filata cheeses develop much better flavor than fast ones.

Fast pasta filata cheeses have acid, in the form of lemon juice, vinegar, or citric acid, added to them, and thus retain milk's natural sweetness. Though this makes for a sweeter cheese, people with lactose intolerance may find difficulty digesting fast pasta filata cheeses. As well, since they don't involve the cultivation of bacterial cultures, fast pasta filata cheeses cannot be preserved well in brine, and should not be aged.

A hybrid of fast and slow pasta filata cheeses can be made by allowing raw milk to ferment naturally on its own for a day at room temperature (but before it thickens into clabber) until it develops enough acidity so that, when set with rennet like a fast pasta filata cheese, it has sufficient acidity to stretch when heated. A braided string cheese of this sort is made by the Doukhobors, a nonmaterial and nonviolent religious group from Russia that fled persecution in the 19th century and settled in the Kootenay Mountains of British Columbia. It is likely that this cheese, which resembles a well-loved Russian cheese known as Chechil (similar to Majdouli described in the recipe below), was taken along with the Doukhobors as they emigrated to North America.

Preserving Pasta Filata Cheeses

Slow pasta filata cheeses are best kept in a chilled, salty brine made of their own whey. A 7 percent salt brine (¼ cup of salt per quart of whey—that's 60 mL salt per L) can be prepared with the fermented whey leftover from a slow pasta filata cheese. And the cheeses can be kept in the salty brine in a refrigerator for up to 1 week.

If the brine and cheeses are unbalanced, minerally or acidy, the cheese may melt into the whey. This can be observed quite clearly if fast pasta filata cheeses are left in a brine made of their whey and water. The fast pasta filata method results in a minerally unbalanced cheese that quickly loses its firmness in a salty brine and melts into the whey.

Fast pasta filata cheeses are best preserved dry. However, they will not keep as well as slow pasta filata cheeses preserved in brine—another good reason to make your mozzarella slow!

- RECIPE -
SLOW MOZZARELLA

This recipe makes true mozzarella: no shortcuts, no shortcomings. Traditional mozzarella is made slowly, patiently, and the results speak volumes of the worth of waiting. When compared with fast mozzarella, well, there really is no comparison.

Made with whole-fat, unprocessed cows' or goats' milk, this cheese sings; made with buffalo milk, the results are symphonic. In the southern Mediterranean climate of Italy, from whence this cheese hails, water buffalo are prized for the richness and whiteness of their milk that makes a richer, whiter mozzarella: Mozzarella di Bufala Campana is a PDO designation that protects the cherished Italian tradition of making mozzarella with buffalo milk.

Handmade mozzarella may be the pinnacle of fresh cheeses. Soft and supple, textured yet juicy, there are many layers to a well-made mozzarella. Wrapped upon itself in numerous leaves, the flaky texture of a hand-pulled mozzarella can be compared to a buttery croissant. Bocconcini, "little mouthfuls" in Italian, is a variation on mozzarella, made by breaking small, bite-sized balls of cheese off the melted and stretched curd.

To make mozzarella, slowly, starter culture is added to the warm milk; after a period of incubating, rennet is added. Once clean break is achieved, the curd is cut and stirred, then strained according to the basic rennet curd recipe in chapter 13. After the curd has knit together, it is put back in its whey to ferment and develop its acidity. As the curd becomes more acidic, the cheesemaker tests the acidity every hour: A small amount of cheese is put in hot water, and if the cheese spins a fine fiber, the curd is ready. If the curd is not yet ready, the cheesemaker waits and tries again an hour later. It usually takes between 8 and 12 hours of fermentation for the curd to spin. I often leave the curd to ferment overnight and finish my mozzarella the next day.

When the curd spins, the cheesemaker warms a pot of lightly salted water to a hot temperature (around 150°F/65°C). As the water warms, a light salt brine is prepared from the leftover whey. The curd is submerged in the hot-water bath, and as it warms it takes on a molten texture. The hot curd is carefully removed from the water, stretched and folded, and resubmerged in its bath until its texture is uniform. The cheesemakers then give the mozzarella its shape and delicate flaky texture by stretching it thin, then rolling the cheese upon itself into a small ball. The mozza balls are placed into cold water to firm them up and keep their round shape, then transferred to a salty brine to develop more flavor and preserve them.

INGREDIENTS

1 gallon (4 L) good milk
¼ cup (60 mL) kefir or active whey
Regular dose rennet (I use ¼ tablet WalcoRen)
¼ cup (60 mL) good salt

EQUIPMENT

1-gallon (4-L) pot
Wooden spoon
Large slotted spoon
3 Camembert-sized cheese forms about 4 inches
    (10 cm) across and 3 inches (8 cm) deep
Cheese knife
Large bowl

TIME FRAME

8–12 hours

YIELD

Makes 3 medium-sized mozzarella balls, about ¾
    pound (340 g) cheese

To make slow mozzarella, leave a basic rennet cheese to ferment in its whey for several hours; submerge the cheese in hot water until it stretches; and then roll the stretched cheese into a ball.

Technique

Warm milk to 90°F (32°C), or baby-bottle-warm.

Add starter culture, either kefir or whey. Mix the culture in well, then cover the pot and keep warm for 1 hour.

Add rennet, and incubate 1 hour: Measure out the appropriate amount of rennet for your milk. Dissolve it in ¼ cup (60 mL) water, then gently mix it into your warm milk. Cover the pot, and keep warm for another hour.

Check for clean break to determine if the curd is ready to proceed.

Cut the curds to ¾-inch (2-cm) pieces by making three series of cuts—one vertical, another vertical but perpendicular to first cut, and the third on an angle close to horizontal. Stir the curds lightly, and cut any large curds to size.

Stir lightly every 5 minutes for 30 to 60 minutes. Warm up the pot slightly, if need be, to maintain a constant 90°F temperature.

Pitch and whey off. Once the curds have developed a poached-egg-like firmness, let them settle in the pot for 5 minutes. Pour off the whey, and reserve it all for fermenting your curds.

Transfer the curds by hand into the cheese forms. Let the curds drain in their forms for 1 hour, until they knit together into a cheese.

Ferment the cheeses in their whey: Remove the cheese from their forms, and submerge them in a potful of their leftover whey. Allow the curds to slowly ferment in the whey at room temperature. Keep the pot covered to keep out flies.

Do a stretch test every hour: Every hour or so, submerge a small piece of curd into hot water. Wait 2 minutes, then examine its stretchiness. If the curd spins a very fine thread when stretched, it is ready for the next stage. If it does not, wait an hour, and try the stretch test again.

Prepare a hot-water bath: Heat ½ gallon (2 L) of water to a hot temperature—around 150°F (66°C).

Prepare a light salt brine by dissolving ¼ cup (60 mL) of salt into 1 quart (1 L) of the whey in a bowl.

Submerge the curds in the hot-water bath. Allow them to warm for 5 to 10 minutes; then, with a slotted spoon, retrieve them one at a time from the water to shape them.

Stretch and roll your mozzarella balls: The hot curd will begin to become silky and plastic. Stretch and knead the curd for a brief moment, and submerge it again in the water to continue warming. Once it is thoroughly plastic, stretch the curd into a long, thin rectangle, roll it onto itself while tucking in the edges, and form the curd into a small, round shape. Stretch the tail of the curd thin and tuck the ball of cheese within to finish.

Seal the mozzarella: Resubmerge the mozzarella in the hot water for a minute, then cup it firmly in your hands to give the cheese its final shape.

Submerge your mozzarella balls in the salty brine to cool. Eat and enjoy while still slightly warm.

Preserve the mozzarella: It can be kept, submerged in its brine, in a refrigerator for up to 1 week.

– RECIPE –
FAST MOZZARELLA, NATURALLY!

Fast mozzarella takes a simple shortcut to achieve in about 1 hour the same miraculous mozzarella stretching that normally takes 8 to 12 hours to achieve. A direct addition of acid to the milk attains the target acidity that is normally reached only through a long, slow fermentation. More technical than traditional, fast mozzarella requires precise temperature controls and ingredient measures.

Most cheesemaking guidebooks suggest using citric acid, a common food additive, to achieve the ideal acidity for making fast mozzarella. The addition of this ingredient, seemingly harmless, raises two concerns. First, it makes a mozzarella that's almost completely tasteless—the citric acid adds no flavor to the cheese, which itself has only a very slight milky flavor. Second, citric acid is often a genetically modified ingredient. Using this GM

ingredient, just like using GM rennet, in turn makes a cheese genetically modified.

The standard method for fast mozzarella stinks of fast-food thoughtlessness. A much more natural way to make it, one that is non-GM and actually has flavor, is to use lemon juice in place of citric acid. What a discovery this was for me when I realized (it's not too much of a stretch!) that lemon juice (or balsamic vinegar!) could replace the citric acid called for in fast mozzarella recipes. It's a perfect stand-in, one that gives this cheese much greater flavor and avoids ingredients of questionable origin.

Lemon juice helps to achieve the target acidity for mozzarella stretching when enough of it is added to the milk. Though the acidity of lemon juice may vary from lemon to lemon, I find that the amount of juice needed holds true regardless of the lemons you use. What proves to be more variable is the acidity of the milk!

Depending on the source and quality of the milk you use for mozzarella-making, you may have to use more or less lemon juice. The standard recipe written here calls for ½ cup of lemon juice per gallon (120 mL per 4 L) of fresh, unprocessed cows' milk. If, lucky you, your milk is water buffalo milk, use slightly more lemon juice, about ⅔ cup per gallon of milk (150 mL per 4 L), as buffalo milk has a higher solids content that buffers against added acidity. The higher solids content also means you'll get more mozzarella—nearly twice the yield of cows' milk. If your milk is a few days old, you will have to add slightly less lemon, as the milk will have already developed some inherent acidity. And depending on the season and the animals' feed, which can both greatly affect milk quality, more or less lemon juice may have to be added.

An unfortunate consequence of directly acidifying the milk to make mozzarella is that this fast mozzarella is not minerally balanced with its whey and does not keep well in a brine: Within a day the cheese's rind will melt into the whey. This fast mozzarella is best left in the brine until cool and firm, then removed and kept dry in a container in the refrigerator for up to 1 week.

INGREDIENTS

1 gallon (4 L) *cold* good milk
½ cup (120 mL) fresh-squeezed lemon juice (or ¼ cup [60 mL] vinegar)
Regular dose rennet (I use ¼ tablet WalcoRen)
Salt

EQUIPMENT

1-gallon (4-L) pot
Wooden spoon
Du-rag or other good cheesecloth
Large strainer or steel colander
Large bowl
Ladle

TIME FRAME

Only 1 hour!

YIELD

Makes about 4 medium-sized mozzarella balls, about ¾ pound (340 g) cheese

TECHNIQUE

Cool your milk. Make sure your milk is cold! If your milk is too warm, the direct addition of acidity will curdle it, and destroy any possibility of mozzarella-making. If you are using fresh milk still warm from the udder, be sure to cool it in the refrigerator before using.

Dilute the lemon juice (or vinegar) in 2 cups (480 mL) of water and slowly pour it into the cold milk, stirring briskly as you pour. Diluting the lemon juice before adding it to the milk ensures a minimum amount of curdling.

Slowly warm the acidic milk, over a low flame, to baby-bottle-warm, 90°F (32°C). Slowly stir the milk as it heats. Too quick a heating, or a lack of stirring, will create unwanted curdling. If the milk gets too hot, it may also curdle, so pay close attention to the temperature.

Add a regular dose of rennet to the warm, acidic milk. Stir the dissolved rennet into the milk very slowly to ensure an even set.

Check for clean break. The high acidity should help achieve a quick, clean break in about 15 minutes.

Cut the curds, in three series of cuts, into ¾-inch (2-cm) pieces. There is no need to wait in between each series of cuts—this acidic curd is very strong.

Slowly stir the curds as they firm so that they do not stick together. Stir occasionally for 15 minutes, keeping the temperature at 90°F, until the curds have the firmness of a poached egg.

Strain the curds: Pour off the whey, and strain the curds into a cheesecloth-lined colander. Let the curds drain and knit together for 30 minutes.

To make fast mozzarella, add lemon juice diluted in water to cold milk; add rennet; and leave the milk to set.

Prepare a hot-water bath: As the curd drains, quickly heat up 2 quarts (2 L) of water to a hot temperature—around 150°F.

Prepare a salt brine by dissolving ¼ cup (60 mL) of salt into 2 quarts (2 L) of cold water.

Do a stretch test: Submerge a small piece of curd into the hot-water bath. Wait 2 minutes, then examine its stretchiness. If the curd spins a very fine thread when stretched, it is ready. If it does not, the mozzarella will not stretch properly! Adjust the amount of lemon juice in the next batch you make, adding slightly more to further acidify the curds.

Cut the pre-mozzarella curd into smaller pieces, each the desired size of a piece of mozzarella.

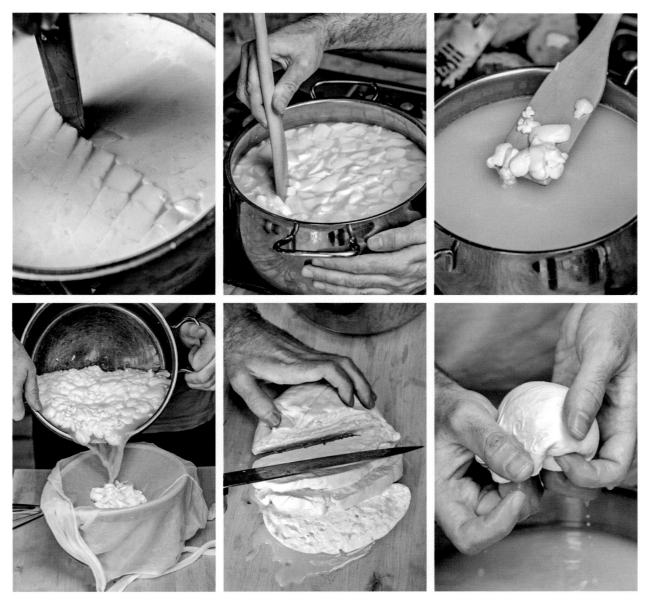

Cut the curd into ¾-inch (2-cm) pieces, and stir the curds for several minutes, until they have the firmness of a poached egg. Strain the curds in cheesecloth until they knit together; cut the curd into thick slices; and finally submerge the slices of curd in hot water and stretch and shape them into mozzarella balls.

Submerge the cut curds in the hot-water bath, using the ladle. Allow them to warm for 5 minutes, then retrieve them with the ladle one at a time from the water to shape them.

Stretch and roll your mozzarella balls: The hot curd will become silky and stretchy. Stretch and knead the curd very briefly, and submerge it again in the hot water. Once it is thoroughly stretchable, pull the curd into a rectangle and roll it onto itself into a small, round shape. Stretch the tail of the curd thin, and tuck the ball of cheese within to finish.

Seal the mozzarella: Resubmerge the mozzarella in the hot water for a moment in order to seal it shut. Cup it firmly in your hands to give the cheese its final shape.

Submerge your mozza balls in the salty water to cool. Eat and enjoy, while still slightly warm.

- RECIPE -

OAXACAN STRING CHEESE

The original string cheese, Oaxacan cheese, aka *Queso Oaxaca*, is the most playful of cheeses. Made of strings upon strings upon strings, this cheese can be pulled apart into smaller strings endlessly.

Queso Oaxaca is usually prepared with milk skimmed of its cream, which gives the cheese a higher protein content and almost infinite stringiness. The cream, which does not contribute plasticity to the curd, is skimmed and reserved for making *crema agria*—homemade sour cream—an essential addition to every Oaxacan meal.

Queso Oaxaca is pulled and kneaded over and over in a particular way to align the strings of curd in one direction. The pulled curd is then shaped into a long ribbon and rolled into a ball. When the cheese is cooled, its strings remain, forming the integral structure of this fine cheese.

INGREDIENTS

As per mozzarella, fast or slow

EQUIPMENT

As per mozzarella, fast or slow

TIME FRAME

1 hour (fast)–12 hours (slow)

YIELD

Makes 1 large Oaxacan string cheese, about ¾ pound (340 g) fresh cheese

TECHNIQUE

Make a batch of pre-mozzarella curd as per slow or fast mozzarella. Be sure that the curd spins well before you continue with Queso Oaxaca.

Prepare a hot-water bath: Warm up ½ gallon (2 L) of water to a hot temperature, around 150°F (65°C).

Prepare a light salt brine by dissolving ¼ cup (60 mL) of salt into 2 quarts (2 L) of the remaining whey.

Warm up the curd in the hot water by cutting it into 1-inch (2.5-cm) slices and leaving these in the hot water for 10 minutes.

Stretch and knead the curd, re-submerging it in the hot water periodically, until it develops an even smoothness and very strong plasticity. Combine all the pieces of curd into one lump.

Pull and fold the curd, pulling until the curd is a foot long (30 cm), then folding it over and over to develop the cheese's strings. Submerge the cheese, as needed, into the hot water to preserve its plasticity.

Pull and fold the curd over and over, working its strings and firming it up, resubmerging it in the hot whey between stretchings.

Roll the curd into a ball: Stretch it out into a long, flat, and even ribbon, then roll it up onto itself as if winding a ball of yarn.

Cool the string cheese by dropping it into a cold-water bath to take away its heat and seal it shut.

To make Oaxacan string cheese, repeatedly stretch and fold a pasta filata cheese to develop its strings; roll the pulled cheese into a ball; and submerge the cheese in a brine.

Brine the Queso Oaxaca: Place the finished cheese into the whey brine to take up some salt and develop some flavor. Slow Oaxacan string cheese can be kept in a brine in the refrigerator for up to 1 week. Fast Oaxacan string cheese should be kept dry in a container in the refrigerator.

– RECIPE –
MAJDOULI

This cheese, relatively unknown in the West, is immensely popular in the Arab world. Wrapped onto itself with an immense twist, this beautiful cheese resembles a skein of silky, white wool.

Because it is usually prepared with the milk of goats or buffalo, two animals common in the Middle East, this cheese is bright and white. Nigella (aka black cumin) seeds, beautifully shaped and jet black, are mixed in with the curd, providing striking contrast and fine flavor to the white cheese.

To make Majdouli, fast or slow mozzarella curd is stretched into a bundle of fibers, then wrapped into a skein. For the curd to respond well to the stretching it must be submerged in extra-hot water, making it soft and supple, and giving it extra plasticity. And the curd must be worked quickly to ensure it does not cool and lose its stretch.

The melted curd is kneaded with nigella seeds, then oiled to help define the strings. The curd is formed into the shape of a donut, and the donut pulled into a large loop. The loop is folded with a twist to make two loops, and the curd stretched once again. The curd is stretched and folded, over and over, until it has become a bundle of cheesy fibers. The round of fibers is twisted in hand, then one end is tucked into the other to hold its spiral, just like a skein of wool.

A certain speed and fluency with the curd is needed to pull off this beautiful cheese—it took me many hours of practice to perfect. Fortunately,

because of the plastic nature of pasta filata cheese, you can work and rework your Majdouli until it's perfect. Watch the pros at work by looking up videos of "Armenian stretch cheese" online.

INGREDIENTS

As per mozzarella, fast or slow
1 tablespoon (15 mL) nigella seeds
Olive oil

EQUIPMENT

As per mozzarella, fast or slow

TIME FRAME

1 hour (fast)–12 hours (slow)

YIELD

Makes 1 Majdouli, about ¾ pound (340 g) fresh cheese

TECHNIQUE

Make a batch of pre-mozzarella curd as per slow mozzarella or fast mozzarella. Be sure that the curd spins well before continuing with Majdouli-making.

Prepare a hot-water bath. Quickly warm up ½ gallon (2 L) of water to a very high temperature, around 180°F (82°C). The water must be very hot to give the curd extra plasticity.

Prepare a salt brine by dissolving ¼ cup (60 mL) of salt into 2 quarts (2 L) of whey.

Warm up the curd in the hot water by cutting it into two pieces and leaving them in the hot water for 5 minutes, until they have developed the ability to stretch through and through. Knead and fold the two curds together, and resubmerge in the hot water until fully plastic.

Fold in the nigella seeds: Retrieve the curd from the pot, place it on a cutting board covered with a sprinkling of nigella seeds, and quickly knead the curd several times to develop its stretch and incorporate the seeds.

Make a cheese donut: Pressing through the middle of the cheese with your finger, form the

To make Majdouli, shape a pasta filata cheese into a donut; stretch the donut into a loop; and fold and stretch the loop over and over to develop the strings of the cheese.

stretchy curd into a donut shape. Splash the donut with olive oil to grease it up.

Stretch the curd: Working quickly, stretch the donut of curd out to arm's length (3 to 4 feet or around 1 meter), whipping the strings as they stretch to ensure an even loop.

Fold the loop in half while giving it a half twist, transforming the single loop into a double loop.

Stretch the double loop out to arm's length again. Plunge the curd into the hot water for a moment to keep it warm and stretchy if need be.

Keep stretching the loop of curd to arm's length and folding it in half, doubling the number of strands each time, until individual strands are less than ⅛ inch (⅓-cm) thick. This takes four or five stretches and folds.

Twist the cheese in your hands: Holding each end of the loop, twist the mass of strings between your hands several times to develop twisted tension.

Tie off the skein: Stuff one end side of the twisted loop of curd through the farthest twist on the other end of the loop. The twisted tension of the skein will thus be held in place.

Place the Majdouli in the salt brine to cool. Slow Majdouli can be kept in a brine in the refrigerator for up to 1 week. Fast Majdouli should be kept dry in a container in the fridge.

CHAPTER FIFTEEN

Feta

Feta is a cheese that is aged by submerging it in a brine made of its whey. Started as basic rennet curds, feta is defined by this brine-aging.

Brine-aged cheeses are most popular in warmer climes, where salty brines preserve cheeses well despite high temperatures and without refrigeration. Brined cheeses similar to feta are made all over Southern Europe, the Middle East, and North Africa and include Turkish white cheese, Palestinian Jibneh Ackawi, and Egyptian Domiati. In these regions cheeses are often preserved in brines along with pickling olives and vegetables; brine-aged cheeses are therefore often referred to as pickled cheeses!

Benefits of Brine-Aging

Submerging a cheese in brine greatly simplifies its aging.

The salty conditions in the brine control the growth of bacteria in the cheese and slow the aging process. The effect of brine-aging feta is similar to that of salt-brining pickles or fermenting sauerkraut: The salty conditions discourage unwanted bacteria and encourage the growth of beneficial cultures that are resistant to the higher salt levels. It is the salt-tolerant lactic acid bacteria that help develop the sourness feta, sauerkraut, and brined pickles are known for.

Brine-aging also helps cheeses be fungus-free. Preserving a cheese submerged in brine prevents it from being exposed to air, which stops fungus from establishing itself.

Brine-aging allows a cheese to be preserved with very little care and attention. Brine-aged cheeses do not need a humidity-controlled aging environment, and they do not need to be washed regularly to keep fungal growth at bay; they just need to be kept cool. But even at higher temperatures a saltier brine will control unwanted bacterial growth that could impact a cheese.

Brine-aging also preserves the texture of a cheese without causing it to dry out or soften. A feta made firm and crumbly will remain firm and crumbly as it ages in its brine; and a feta made soft and creamy will stay soft and creamy.

Developing Feta's Tang

Feta's origins trace back to the traditional cheese-making customs of shepherds in the hills of Greece, to whom packaged starter cultures and commercial enzymes were entirely foreign and unnecessary. Today, however, most fetas are produced on an industrial scale according to standard freeze-dried cheesemaking methods, and their taste, of course, suffers. A more authentic and tangier-tasting cheese can be made with simpler peasant practices. (By using the term *peasant*, I mean no disrespect. I'd prefer the French term *paysan*, a title that is used with honor to describe farmers who practice simple, small-scale, and sustainable agriculture.)

The simpler practices that make a tastier feta are using raw milk, culturing with backslopped whey, renneting with dried lamb or kid stomachs, and a many-months-long curing in wooden barrels of brine. Feta cheeses made in a more modern manner just won't taste the same.

Made with good raw milk, traditional feta has much more flavor in part because of raw milk's indigenous microbiodiversity but also because the sheep and goats that provide the milk are pastured.

Traditional feta-makers start their cheeses by backslopping with whey left over from a previous batch of cheese. By invigorating the culture every time they make their cheeses, they cultivate the diverse raw milk bacteria that help give complexity to their cheeses.

Feta was originally coagulated with cured kid or lamb stomachs as a source of rennet. This unprocessed rennet contained lipase and other enzymes that help feta develop its tangy flavors. Many cheesemaking recipes call for the addition of lipase (yet another ingredient to purchase from cheesemaking supply shops) to help develop the tanginess that feta is known for; however, you can get as good an effect by making feta with raw milk or using kefir as a starter, both of which contain bacterial cultures that produce lipase naturally, eliminating the need to add this enzyme to the milk. Using less processed rennet—in the form of kid rennet paste, or from your own dried vells—intensifies the effect.

Feta is best aged, like wine, in wooden casks filled with brine. The wooden barrels preserve cultures and flavors from previous batches of feta that only improve subsequent batches of cheese aged in them. The flavor of the wood itself also adds nuances to feta, much as it does to wine.

The longer a feta is aged in its brine, the tangier it will get. Fetas can be eaten after only a couple of weeks of aging. But patience rewards those who wait: Brine-aged cheeses develop their best characteristics only after many months.

Making and Using Brines

Feta cheeses are left to age in a light brine made of their whey and salt. Much less salty than the saturated salt brine used to salt cheeses, an aging brine is usually made to be 7 percent salt by weight.

To prepare an aging brine, add to the whey 7 percent of the weight of whey in salt. To help with the math, that works out to roughly 1 cup (240 mL) of salt per gallon (4 L) of whey, or ¼ cup (60 mL) per quart (1 L). Be sure to use good noniodized salt to make the brine: Fine sea salt, kosher salt, or mineral salt all work well. Iodized salt, with its antiseptic iodine, can disrupt the microbial balance in the brine, resulting in the growth of unwanted bacterial cultures that can have adverse effects on an aging feta.

Make the aging brine with leftover whey from the cheese that's to be aged. And leave the whey to ferment and develop its acidity along with the cheese until it is ready to be submerged in the brine. If the cheese and the brine have different acidity levels, the aging cheeses can develop slimy rinds.

If you wish to preserve brined cheeses for many months, or at higher temperatures, consider doubling the salt content of the brine. The more salt there is in the brine, the less microbial growth there will be within a cheese, and the longer the cheese will keep.

Preparing Cheeses for Brining

Cheeses must be properly salted and air-dried before being submerged in their brines. Salting and drying help to keep feta cheeses firm as they age.

Salting the surfaces of the fetas, followed by a daylong period of air-drying, is essential for the development of strong rinds that prevent the cheeses from melting into their brines. If the feta is cut into pieces before submerging in its brine, all of its surfaces must be salted, then the pieces left to drain and dry to be sure that all sides have strong rinds.

Using the Aging Brine

Store the brine in a nonreactive vessel. Glass is preferred, and plastic works, too, but try to avoid metallic containers, which may corrode in the salty brine. If you're making big batches of feta, large ceramic crocks or oak casks will give the finest results.

Brines should be kept cool, with temperatures less than 50°F (10°C): The cooler the temperature, the slower the feta will age and the longer it will keep. Brines can be kept cool in the refrigerator, in a cellar, in an unfinished basement, or in a cold room, but there is no need to keep the brine-aging cheeses in a humid environment as with other aged cheeses—the brine keeps the cheeses as humid as they need to be!

Keep the aging brine covered to limit evaporation, but do not use metallic lids to cover the fetas: When exposed to the salty and acidic brine, the metal may rust and can impart metallic flavors to the cheese. Be wary as well of the possibility of the cheese giving off gas, and open up the lid every so often to release any built-up pressure.

Cheeses must be kept submerged within their brine. A cheese exposed to air will be susceptible to fungal growth. If the cheeses float, place a stone or other weight atop to keep them under the whey. If you suffer from floating fetas, try making your cheese with fresher milk to limit the formation of gas bubbles within the cheese.

- RECIPE -
FETA

True feta (the name is now controlled under European Union trade agreements) can only be made with the milk of goats and sheep that graze the Greek hilltops.

Naturally more flavorful than cows' milk, goats' milk and sheep's milk make a feta with much more complexity. Goats' and sheep's milk also give feta a bright white color and richer texture, while feta made with cows' milk will have a creamier color and firmer texture. Nevertheless, a "feta-style" cheese, made with good local cows' milk, is still sure to please!

Feta starts as a basic rennet cheese, made with fresh milk, culture, and rennet. The curd is cut then stirred until it has a slight firmness like other rennet cheeses. Once the curds are ready, however, they are salted before forming to pull out their whey and develop the crumbliness that fetas are known for. After salting, the curds are pressed firm, then cut into slices, and their surfaces are salted. The slices of cheese are air-dried and, finally, submerged in a light salt brine to age.

INGREDIENTS

1 gallon (4 L) good cows', goats', or sheep's milk
¼ cup (60 mL) active kefir or whey
Regular dose rennet (I use ¼ tablet WalcoRen calf rennet for 1 gallon milk)
Good salt

EQUIPMENT

1-gallon (4-L) stainless pot
Wooden spoon
Long-bladed knife
Spaghetti colander
Large stainless bowl
1 quart-sized (Litre-sized) cheese form and follower for pressing
Draining table setup
Cheesecloth
1-quart (1-L) glass jar for brine-aging
Refrigerator for aging

TIME FRAME

4 hours to make; 2 weeks minimum to age

YIELD

Makes 1 pound (450 g) feta

TECHNIQUE

Slowly warm the milk to 90°F (32°C), baby-bottle-warm.

Add the active whey or kefir, and mix it in lightly.

Incubate 1 hour: Cover the pot, and keep the milk at 90°F for 1 hour to encourage the bacterial cultures to ripen it.

Add the rennet: Dissolve the regular dose of rennet in ¼ cup (60 mL) water, and gently mix it into the milk.

Incubate 1 hour: Cover the pot and keep the milk at 90°F for another hour to encourage the rennet to set the curd.

Check for clean break. Proceed to cut the curd once it has set.

Cut the curd into ¾-inch (2-cm) pieces in a series of three cuts, waiting a few minutes between each series of cuts to let the curds heal.

Slowly stir the curds for 30 to 60 minutes, every 5 minutes, until they develop a poached-egg-like firmness.

Pitch, then whey off: Let the curds settle for 5 minutes, then pour off the whey, reserving 1 quart (1 L) for making an aging brine.

Salt the curds: Strain the curds from the pot into a colander placed over a bowl. Sprinkle 2 tablespoons (30 mL) of salt over them, and mix the salt around with your hands. Let the salt pull moisture out of the curds for 10 minutes, stirring the curds a few times to prevent them from sticking together.

Set up the cheese press atop the draining table. Lay cheesecloth in the form to help hold the curds together as they have their first pressings.

Place the curds into the form of the press, filling it to within no more than an inch (2.5 cm) of its top.

Place the follower atop the cheese, half-filled with warm whey, to serve as a press.

Prepare a light salt brine: While the cheese is pressing, prepare a light salt aging brine with 7 percent salt by dissolving ¼ cup (60 mL) of salt into a quart (1 L) of still-warm whey. Leave the brine to ferment at room temperature, covered with a cloth, until the feta is salted and air-dried and ready to be submerged.

Flip the cheese several times while pressing: Take the cheese out of the form with the aid of the cheesecloth, then rewrap it upside down. Place the cheese back in its form upside down, so that it can be pressed on the other side, and pour more whey into the follower to increase the pressing weight. Continue pressing the cheese, flipping it every 10 minutes as it firms.

Take the cheese out of the press once it has cooled. After the cheese has cooled, the pressing will lose its effectiveness. The cheese can be left to rest in cheesecloth in its form overnight.

To make feta, first make basic rennet curds; drain them in a colander, and salt them; then press them into a cheese before brining.

Cut and salt the cheese: Remove the cheese from its form, and cut it into pieces that will more easily fit in your aging vessel. Apply salt to the surfaces of each piece to ensure that they develop strong rinds.

Air-dry the cheeses on a draining rack at room temperature for a day or two. Flip them once or twice to help them dry evenly.

Brine-age the cheeses: Place the cheeses into the light salt brine in a jar, and leave them to age in a cool place with temperatures less than 50°F (10°C) for at least 2 weeks. Keep the cheeses submerged, if need be, by placing a weight atop them or by wedging them into the jar with a few wooden sticks. Cover the jar with a cloth lid.

Check on the cheeses once a week as they age, to be sure that they are submerged in their brine. Feta will be ready in 2 weeks, but best results come after 2 months of aging.

- RECIPE -
CREAMY FETA

Some folks like their fetas firm and crumbly; others prefer theirs soft and creamy. A softer version of feta can be made using a lactic curd (essentially chèvre, made in chapter 11) instead of a firmer full-rennet curd as in the feta recipe. Commonly called Bulgarian feta, this cheese is now labeled as Bulgarian white cheese because of the PDO protecting the name *feta*.

Bulgarian feta uses a lighter dose of rennet and a longer fermentation time to make a creamier and more flavorful curd. The soft curd is drained and formed, salted, and air-dried into a shaped chèvre. The round of cheese is then brine-aged just like a firmer feta. Try the two fetas side by side to see which version you prefer.

INGREDIENTS

1 gallon (4 L) good milk
¼ cup (60 mL) active kefir or whey
¼ dose rennet (1 use ¹⁄₁₆ tablet WalcoRen calf rennet)
Good salt

EQUIPMENT

1-gallon (4-L) cheesemaking pot
Slotted spoon
6 cylindrical goats' cheese forms
Draining table
1-quart (1-L) glass jar with lid for brine-aging
Refrigerator for aging

TIME FRAME

1 hour over 3 days to make; 2 weeks–2 months to age

YIELD

Makes 6 small Creamy Feta cheeses

TECHNIQUE

Slowly warm the milk to 90°F (32°C), baby-bottle-warm, then turn the heat off.

Add the kefir or whey to the pot along with the small dose of rennet dissolved in ¼ cup (60 mL) water. Gently stir the milk to incorporate the rennet and culture.

Cover the pot, and let it ferment at room temperature for 24 hours. During this time, the curd will acidify, firm up into soft curd with a yogurtlike consistency and sink under its whey to the bottom of the pot.

Ladle the curds into the forms: With the slotted spoon, transfer the curds from the pot to the forms atop the draining table. Fill the forms right to their brims with curd.

Prepare a light salt brine: As the curds are draining, prepare a 7 percent salt brine with 1 quart (1 L) of whey in a large glass jar. Dissolve ¼ cup (60 mL) of salt in the whey, then leave the brine at room temperature to ferment until the cheese is ready to be brine-aged.

Drain 24 hours. Flip the young cheeses at some time after 12 hours of draining, whenever is convenient. The cheeses will be very soft, so handle them gently.

Salt the cheeses by applying 1 teaspoon (5 mL) of salt around the surfaces of each cheese.

Air-dry the cheeses on a draining mat at room temperature for 24 hours, flipping them once or twice to ensure an even drying. When the cheeses are dried to the touch, they can be brine-aged.

Brine-age the cheeses: Place the cheeses into the light salt brine in a jar, and leave them to age in a cool place with a temperature less than 50°F (10°C). Keep the cheeses submerged, if need be, by placing a weight atop them. The Creamy Feta will be ready in 2 weeks but can age up to 2 months.

To make Creamy Feta, ladle soft chèvre curd into forms; salt and air-dry the firmed cheese; and submerge the cheese in whey brine to age.

CHAPTER SIXTEEN

White-Rinded Cheeses

The beautiful white rinds of Camembert and Brie are the manifestation of a fungus that grows on these cheeses. Appearing to the naked eye as a velvety white coat, the rinds are, in fact, a dense mat of interwoven mycelia of a fungus that's in the midst of consuming the cheese.

It is fungus that gives white-rinded cheeses their best qualities. Their molten interiors are made possible by fungus, as are their intoxicating aromas and, of course, their mushroomy flavors.

White-rinded cheeses can be made from every different type of curd. Regular rennet curds can be ripened with fungus into a Brie or Camembert (recipes on page 189). Softer lactic curd or even yogurt cheese can be ripened into a white-rinded lactic cheese such as Crottin (as seen in chapter 12, Aged Chèvre Cheeses). And Alpine cheese curds can be ripened into a white-rinded Tomme (see chapter 19). Each of these cheeses starts off as a fresh cheese, but particular fungal cultures encouraged by the cheesemaker come to define their development as they age.

How Does Your Fungus Grow?

Two cultures, *Penicillium candidum* and *Geotrichum candidum*, are the fungi of choice for making white-rinded cheeses, though many species of yeasts are also integral to the development of healthy white rinds.

When cheesemakers create a white-rinded cheese, they either rely on indigenous fungal cultures present in their raw milk or intentionally introduce the fungus to it. Either way, the milk begins its transformation into cheese with the appropriate fungal spores already in it.

As the milk is coagulated into curd and the curd knits into cheese, the fungal spores come to life. Two spores find each other and germinate in a sort of reproductive dance that's not too different from our own. The germinated spores then send out their mycelial roots, looking for food to feed on—and they find abundant nourishment within the cheese.

When the cheese is put in its cave to age, the fungal mycelium feeds on the lactose sugar and lactic acid, as well as casein and other proteins present in the flesh of the cheese. The fungus breaks down these elements and produces aromatic compounds that give white-rinded cheeses their characteristic flavors.

As the fungus feeds, it also produces wastes that reduce the acidity of the cheese, affecting the way the calcium of the cheese interacts with the proteins. The calcium removes itself from the protein chains, weakening the proteins' structure and liquefying the cheese: The oozing centers of white-rinded cheeses are a consequence of the cheeses' declining acidity.

The moisture-rich conditions of the cheese cave sustain the growth of the fungus. At the surface of the cheese, where the air it needs to breathe is most plentiful, the fungus produces its thickest growth. Layer upon layer of white mycelial filaments begin to cover the rinds of these cheeses, and as the fungus matures, the mycelium produces its fungal fruiting bodies. Microscopic mushrooms bloom on the surface of the cheese, and as the fruiting bodies ripen, they send off their spores into the world in search of another cheese to grow on.

Geotrichum vs. Penicillium

Both *Geotrichum candidum* and *Penicillium candidum* grow white rinds on cheeses. The Latin name *candidum* refers to the white color of both these fungal cultures; however, *Geotrichum* gives cheeses a cream-colored rind, whereas *Penicillium* grows bright white.

P. candidum tends to grow a thick mat of mycelium atop a ripening cheese, while *Geotrichum* tends to grow shorter and has a more powdery appearance.

Geotrichum, more than *P. candidum*, also influences the texture of the rinds.

The softest cheeses, especially aged chèvres such as Crottin, tend to grow rinds that are wrinkles upon wrinkles if they are ripened with *Geotrichum*. Semifirm cheeses such as Camembert tend to have more rounded and rippled surfaces. And firm cheeses, such as Alpine cheeses, get a fine dusting of white. The same cheeses, if aged with *P. candidum*, grow flat, undistinguished white rinds.

Both *Geotrichum* and *Penicillium candidum* soften cheeses in a similar way as they ripen, and there is also no significant taste difference between cheeses ripened with the two fungi. In my opinion, though, *Geotrichum candidum* is more desirable: It makes a more beautiful cheese, it's stronger and more resilient than *Penicillium candidum*, and, most important, it's much more easily kept at home!

Sources of Spores

Contemporary cheesemakers add laboratory-raised *Penicillium candidum* or *Geotrichum candidum* spores to their Camemberts. Whenever they wish to start a batch of white-rinded cheese, they sprinkle a prescribed dose of spores atop the milk at the beginning of the cheesemaking process.

But you need not resort to packaged culture to grow white rinds on your cheeses: Appropriate fungal spores can be found in both raw milk and kefir! Raw milk, with indigenous *Geotrichum*, is the traditional source of white fungus, and raw milk cheeses ripened in the right conditions will bloom beautiful white rinds without any added spores. Kefir also contains *Geotrichum candidum*: Its diverse fungal and bacterial cultures make it an excellent starter and ripening culture for making white-rinded cheeses.

DVI Spores

Cheesemakers who use DVI fungal cultures are only adding one single fungal species to their cheeses, be it *Penicillium candidum* or *Geotrichum candidum*. This simplification creates a microbiological imbalance in the cheese that can result in weaknesses in its ecology, because many different species, including yeasts, are needed to establish healthy white rinds.

DVI fungal cultures make cheeses that may be more susceptible to foreign fungal growth, as the single species they introduce doesn't have the community of microorganisms it needs to thrive. As a result of the use of DVI fungi, contemporary cheesemakers must take extra precautions to keep their cheesemaking operations sanitized to reduce the possibility of microbiological contamination. A more relaxed, less sterile approach to cheesemaking can be achieved through using the indigenous fungal cultures found in raw milk and kefir.

Raw Milk and White Fungus

Among the many diverse beneficial bacterial cultures in raw milk are many species of yeast and fungi that help to establish white rinds on cheeses. Present as spores within the milk, these indigenous fungal cultures, which include *Geotrichum candidum*, are all a cheesemaker needs to make cheeses white. Indeed, traditional cheesemakers used nothing but the indigenous fungus of raw milk to encourage the growth of white rinds, and their results were likely better than those of contemporary cheesemakers.

Most cheesemakers today shy away from making soft white-rinded cheeses with raw milk because of limitations on the sale of raw milk cheeses aged less than 60 days. Unfortunately, most small, soft, white-rinded cheeses (such as Camembert, Crottin, and Saint-Marcellin) are not meant to be aged that long, and therefore cannot be sold raw! Quebec, a Canadian province with a deep and distinct cheesemaking culture, has enacted special regulations allowing raw milk cheeses to be sold at less than 60 days, specifically because of the uniqueness of these small, white-rinded raw milk cheeses that would not be the same if made with pasteurized milk.

The common belief is that soft white-rinded cheeses do not sustain a low enough pH (a high enough acidity) for long enough to eliminate the threat of pathogenic bacteria in raw milk, unlike longer-aged harder cheeses. However, the fungal cultures that ripen these soft cheeses likely keep pathogenic bacteria in check through other means: Fungal cultures produce antimicrobial compounds that limit the growth of pathogenic bacteria. Are cheese-ripening fungi not known for their antibiotic effects?

There are plenty of reasons to make white-rinded cheeses with raw milk—most important, they taste better. And if the raw milk is produced in a way that is safe to drink (see chapter 2), then that raw milk should to be safe to turn into cheese. Assuming that cheese made with *bad* raw milk, but aged 60 days, is safe to eat, but limiting the eating of cheeses made with *good* raw milk aged less than 60 days is a dishonorable and destructive regulation that limits traditional cheesemaking practices.

Using Kefir as a Source of White Fungus

Keeping your own white fungal culture is as simple as keeping kefir grains.

Because they contain indigenous populations of *Geotrichum candidum*, kefir grains can be used to establish white rinds on Camembert much the way they can be used to establish white rinds on aged chèvre cheeses. If you're making pasteurized milk cheeses, using kefir as a starter can help reestablish the microbiological diversity that the cheeses need to develop a healthy white coat naturally. Even when you're using raw milk, kefir can help ensure that *Geotrichum* growth gets off to a head start ahead of other unwanted fungi.

The native populations of *Geotrichum candidum* in kefir grains can be observed if kefir is left out at room temperature to ferment, undisturbed, in an unsealed jar (leaving the jar sealed limits oxygen, which restricts fungal growth). After just 2 days, a

Three jars, 3 weeks old, showing the natural progression of, from left to right, raw milk, pasteurized milk, and pasteurized milk inoculated with kefir. Only raw milk and kefir grow wrinkled coats of fungus because of their indigenous *Geotrichum* cultures.

slight fuzziness will begin to cover the surface of the kefir. After 1 week, the fungal growth on the surface of the kefir will be unmistakably *Geotrichum*: pure white, powdery growth, with slight ripples reminiscent of the rinds of *Geotrichum*-ripened cheeses!

Raw milk, if left to ferment in a similar manner, will also show signs of its native *Geotrichum* growth, though it will take longer to establish than kefir. Pasteurized milk, however, will remain fungus-free.

Encouraging White Growth

Once the right fungi are introduced into the cheeses, however, it is not a sure bet that the cheeses will grow white rinds: You still have to create the right conditions for those fungal cultures to thrive.

A cheese with the appropriate fungal cultures will only grow a white rind if that cheese is tended to in just the right way. And a cheese can easily be dissuaded from becoming a white-rinded cheese if it is subjected to conditions that are conducive to the growth of other ripening cultures; for instance, if a young Camembert isn't treated right, it can evolve into a surface-ripened blue cheese, or a stinky washed-rind cheese!

The Right Conditions Encourage White Growth

Salting cheeses well helps to ensure white fungal growth. While *Penicillium* and *Geotrichum* are tolerant of mildly salty conditions, other species of wild fungus are less tolerant of salt and are easily discouraged by a proper salting (see chapter 5). A good salting will help keep yellow, black, and "cat fur" fungus from growing on white-rinded cheeses.

Keeping the temperature of the cheese cave around 50°F (10°C) also encourages white fungal growth. Temperatures too low, such as those found in a household refrigerator, can discourage the growth of *Geotrichum* fungus, and encourage wild *Penicillium roqueforti*, which is more tolerant of cold temperatures.

Keeping the humidity of the cheese cave at 90 percent also helps with the development of white rinds. Too little humidity can limit the growth of fungus, while too much can also hamper fungal growth: Excess humidity may promote the growth of *Brevibacterium linens* bacteria that can cause a white-rinded cheese to ripen into a washed-rind cheese.

White-rinded cheeses that aren't able to breathe can also become washed-rind cheeses. *Geotrichum* thrives on cheeses exposed to air; but if airflow is restricted, the fungal cultures can be overwhelmed by other microbiological regimes. If, for example, a young Camembert cheese is tightly wrapped in plastic to age, the moisture trapped by the plastic can cause the cheese to ripen with *B. linens*.

Isolation from Penicillium roqueforti

A white cheese that's contaminated with *Penicillium roqueforti* spores can evolve into a blue cheese. The presence of small numbers of *P. roqueforti* spores can result in small blue spots on the rinds of Camemberts; more abundant *P. roqueforti* contamination can overtake the development of *Geotrichum* and can give a Camembert a bad case of the blues.

Commercial cheesemakers have zero tolerance for foreign fungi on their white-rinded cheeses, and even a mild amount of *Penicillium roqueforti* growth can cause them to dispose of an entire batch. Many cheesemakers therefore go to great lengths to prevent *P. roqueforti* from getting a foothold on their cheeses. Cheesemakers take extra care to ensure complete sterility in their cheesemaking operation (and even avoid raw milk) to prevent *P. roqueforti* contamination. And many find that they have better success with their white-rinded cheeses if they build separate cheesemaking facilities for their blues!

Thorough cleanliness, without sterilization, can be enough to keep *P. roqueforti* at bay, however, if the rinds of cheeses are washed with whey for their first week of aging. Washing the rinds with salty whey can set back *P. roqueforti* and create conditions that favor the growth of *Geotrichum*.

Washing Rinds to Encourage Geotrichum

Much like weeding a garden to encourage the desired plants to grow, washing the rinds of cheeses

Washing a young Camembert's rinds encourages the development of white *Geotrichum* rinds.

with whey early in their development stops unwanted fungal growth and sets the stage for milk's native *Geotrichum* culture to thrive. Regular rind washings for the first week of a cheese's aging will limit growth of *Penicillium roqueforti* and other fungi that can develop on rinds but will not cause the pink/orange *B. linens* growth that defines washed-rind cheeses that are washed for several weeks (see chapter 18). Once the rind washing has stopped, the *Geotrichum* will have established a foothold on the cheese and will grow a pure white rind.

I make white cheeses and blue cheeses with the same unsterilized tools, and I ripen blue cheeses alongside white cheeses in the same cheese cave without cross-contamination. By handling my cheeses in the right way, I encourage the right fungal cultures to grow, and once *Geotrichum candidum* has established itself on a cheese, it keeps other fungal cultures at bay.

- RECIPE -

BRIE AND CAMEMBERT

Camembert and Brie are made from basic rennet curds, as explored in chapter 13. The cheeses are made with raw milk, started with an active bacterial culture such as kefir, and renneted with a normal dose of rennet. The curds are cut into ¾-inch (2-cm) cubes and stirred until firm, then ladled into appropriately sized forms. The formed cheeses are salted, air-dried, and placed into a cheese cave to age. Rind washing for the first week of the cheese's aging helps keep unwanted fungal growth at bay, and encourages the development of the cheese's white

rind. A further month or two of aging allows the fungus to ripen the cheese to a molten interior.

Brie is a distinct cheese from Camembert, but both are made in a similar way. In North America we hardly distinguish between the two, and many cheesemakers make both Brie and Camembert according to nearly identical recipes. But if you find yourself in France don't confuse the two—you might find yourself deported!

The main considerations that distinguish Brie and Camembert are their size and the terroir, both of which influence the way these cheeses develop. Bries are considerably larger than Camemberts; the size changes the relationship between surface and interior, and affects the way each cheese ages. Camemberts tend to ripen more quickly and become more molten because the small cheese's development is dominated by surface-ripening effects. Bries ripen more slowly, because their large size limits the effects of surface ripening. Bries are often aged 3 to 4 months, whereas Camemberts are typically eaten after 1 or 2: Brie's longer ripening time also allows it to take on much more complex flavors than Camembert.

The flavors of true Bries and Camemberts are representative of the terroir of their place of origin. The two cheeses come from two distinct regions of France: The unique taste of the cheeses made in each region is defined by the nature of the milk that comes from the particular breeds of cows feeding on the distinct plants that grow in the specific soils of each region. Add to that the effect on the ripening cheeses of the distinct raw milk bacteria and fungi from each region, and Brie and Camembert each take on a life of their own.

Bries and Camemberts made outside of their places of origin just won't taste the same. Those made industrially, with milk from cows that have never tasted a blade of fresh grass, and whose milk is pasteurized and recultured with freeze-dried, laboratory-raised bacterial and fungal cultures, will be particularly bland. However, locally made raw milk Bries and Camemberts, produced from raw pastured cows' milk, can taste significantly better than industrial versions of the cheese made in the original cantons of Brie and Camembert. I believe these cheeses are well worth making in any terroir.

INGREDIENTS

1 gallon (4 L) good cows' milk
¼ cup (60 mL) kefir or active whey
Regular dose rennet (I use ¼ tablet WalcoRen rennet for 1 gallon milk)
Salt

EQUIPMENT

1-gallon (4-L) stainless pot
Wooden spoon
Long-bladed knife

To make Camembert, strain rennet curds into a form; salt and air-dry the cheese; and wash its rind for a week as it ages to encourage the growth of white fungus.

1 Brie-sized (7 inches [18 cm] across) cheese form or 3 Camembert-sized (4 inches [10 cm] across) forms
Draining rack setup
Cheese cave at 50°F (10°C) and 90 percent humidity

TIME FRAME

4 hours to make; 1–2 months to age for Camembert; 3–4 months to age for Brie

YIELD

Makes 1 Brie or 3 smaller Camemberts

TECHNIQUE

Slowly warm the milk to 90°F (32°C), baby-bottle-warm.

Add the active kefir or whey, and mix it in lightly. The kefir contains both the bacterial starter culture and the fungal ripening culture. Raw milk contains *Geotrichum*, as does kefir; if you use pasteurized milk and do not add kefir, *Geotrichum* may not bloom on the rinds of the cheese.

Incubate 1 hour: Cover the pot, and keep the milk at 90°F for 1 hour to encourage the bacterial cultures to ripen the milk.

Add the rennet: Dissolve the regular dose of rennet in ¼ cup (60 mL) water, and gently mix it into the milk.

Incubate 1 hour: Cover the pot and keep the milk at 90°F for another hour to encourage the rennet to set the curd.

Check for a clean break and a nice satisfying *pop!* Proceed to cut the curd if it has set.

Cut the curd, in three series of cuts, into ¾-inch (2-cm) pieces. Wait a few minutes between each series of cuts to let the curds heal.

Keep the pot at 90°F while you gently stir the curds for 30 to 60 minutes, to encourage them to firm up. Proceed to whey off once they have the firmness of a well-poached egg.

Whey off: Pour the whey off the pot, and reserve 1 quart (1 L) for preparing a washing brine.

Strain the curds, by hand, into the larger Brie form or smaller Camembert forms. Fill the forms up to the top.

Prepare a washing brine by dissolving 1 tablespoon (15 mL) salt in 1 quart (1 L) of whey. Keep this brine in the refrigerator for washing the rinds of your cheeses.

Form the cheeses for 24 hours: Place the filled forms on a draining rack and let the cheeses drain for 24 hours. Flip the cheeses once or twice during their draining period to give them an even shape. Remove the cheeses from the forms once they are firm, and proceed to salting.

Salt the cheeses: Apply around 1 tablespoon (15 mL) of salt to the surface of the Brie, or 1 teaspoon (5 mL) to the surface of each Camembert. Be sure to spread the salt around the entire cheese, and apply more if some falls off.

Air-dry the cheeses: Let the cheeses drain for at least 24 hours. Flip them once or twice as they dry. When the cheeses are dry to the touch, they are ready to be put into the cheese cave.

Place the cheeses into your cheese cave to age.

Wash the rinds: During the first week of aging, wipe down the rinds of the cheeses every other day with a cheesecloth dipped in the washing brine. Smearing the cheeses' rinds sets back unwanted fungal growth, and gives the *Geotrichum* a chance to thrive.

Check up on the cheeses twice weekly to be sure that the aging conditions are just right. Flip the cheese every time you check on it so that it doesn't stick to its mats. After a month of aging, once the fungal coat is well established, Camemberts and Bries can be wrapped in cheese wrappings and left to age in a refrigerator. Camemberts will be ready in less than 2 months; a Brie will be ready in 3.

CHAPTER SEVENTEEN

Blue Cheeses

Just as in the fabrication of white bloomy-rinded cheeses, cheesemakers intentionally add spores of a specific fungal species to create their blue cheeses. The fungus responsible for making blue cheeses blue is *Penicillium roqueforti*, named for the French town of Roquefort, home to the world's most famous blue cheese.

Penicillium roqueforti

Penicillium roqueforti consumes cheese. When *P. roqueforti* spores find their way into a cheese, they come to life. They spread out their mycelial roots looking for food to eat and air to breathe. Using the cheese as its source of energy and nutrition, the fungus produces waste products that slowly transform the cheese. The by-products of its metabolism change the acidity of the cheese, which, along with the digestive enzymes the fungus produces, begins to soften its flesh.

Penicillium roqueforti protects cheeses. The fungus produces mycotoxins that restrict the growth of other fungal and bacterial cultures. These toxins are similar to the wonder drug penicillin, produced by a closely related *Penicillium* fungus, and have comparable antibacterial properties. These mycotoxins also contribute to blue cheese's distinct flavors, and sometimes even make your tongue numb!

Penicillium roqueforti makes cheese blue. At the surface of the cheese, where it finds plentiful air, the fungal culture thrives. Its mycelium begins to form a dense mat of fibers, and out of that growth the fungus produces its microscopic fruiting bodies— tiny mushrooms that send its spores out into the world. Visible as a soft greenish-blue coat on the cheese, it is the spores produced by the fungal growth that give blue cheese its color.

Understandably, not everyone appreciates a ripe blue cheese; a strong blue cheese is, after all, a cheese in an advanced state of decomposition. That rich ammonia smell that wafts up from such delectable cheeses is indeed the odor of death, decay, and degradation!

Blue Veins

Penicillium roqueforti's blue growth only appears where the fungus is exposed to air. Consequently, fungal growth is naturally thickest at the rind. But most blue cheeses don't show much fungal growth on their rinds; their blue color is most vivid on their interior where *Penicillium roqueforti* manifests itself in the form of blue veins.

Cheese eaters generally do not appreciate the blue, green, and sometimes brown fungal growth on the surfaces of *P. roqueforti*–ripened cheeses. Many cheesemakers therefore limit the growth of *P. roqueforti* by regularly washing their cheeses' surfaces, preventing the fungus from establishing itself there. Turophiles prefer their blue on the interior of the cheese, and cheesemakers encourage blue veining by piercing their blue cheeses with skewers to introduce air into the flesh of the cheese.

By allowing air to pass to the flesh of the cheese, cheesemakers trick the fungus into producing its spores on the cheese's interior. Cheesemakers will pierce a large blue cheese dozens of times to ensure that air passes into every crack and crevice within, supporting the development of the fungus's blue-green spores. The air allows the fungal culture to thrive where it normally would not, and to produce the cheese's characteristic blue veins.

Cheesemakers will even make their blue cheeses in certain ways so that there are more crevices between the curds, and therefore more numerous veins. Several ways that cheesemakers can encourage veining in their blue cheeses are salting the curds on a draining table before they are placed in the form to ensure that the curds don't knit together too much (as is the case with Gorgonzola, recipe on page 206); encouraging gas-producing microorganisms to thrive in their cheeses, creating spaces for fungal growth (something that happens naturally with raw milk cheeses, and how Roquefort gets its blue); and passing on the pressing, which can eliminate the spaces between curds.

Sources of *Penicillium roqueforti* Spores

Contemporary cheesemakers source their blue fungal culture from packages of freeze-dried *Penicillium roqueforti* spores manufactured by culture houses and sold by cheesemaking suppliers. Whenever they want to start a blue cheese, they open up the package, measure out a precise dose of spores, and either add the culture directly to the milk at the beginning of the cheesemaking process, or sprinkle the spores over the top of the curds before placing them in their form.

But no cheesemaker ever needs to buy *Penicillium roqueforti* fungal spores for making blue cheese, for a ripe piece of blue cheese contains *billions* of them! Until I found a better way (see page 197), I kept a blue fungal culture alive for years in the blue cheeses that I made. Every time I started a batch of blue, I

would take a small piece of a ripe blue cheese with blue veins (a sure sign of the presence of *P. roqueforti* spores), dissolve that piece of cheese in water, and pour the water, with its fungal spores, over the pot of milk to introduce the spores to the cheese.

This recycling of fungal cultures works, although it is generally not practiced by commercial cheesemakers for a number of reasons: There are contamination concerns arising from uncertainty about the purity of the fungal spores sourced from a cheese; there are food safety concerns about integrating batches of cheese and the spread of pathogens; and it's much easier to just open up a package of spores and pour them over the milk!

Penicillium roqueforti: *A History*

Traditional cheesemakers in the Pyrenees region of France didn't have access to freeze-dried fungal spores when they made their Roquefort cheeses. They had a trick up their sleeves, a secret they used to make their trademark blue cheese. To help contextualize their practice, I'll first tell the story of how blue cheese was discovered.

Cheesemakers in the village of Roquefort, where blue cheese was born, tell visitors to their famous cheese caves a story of blue cheese's origins that goes a little something like this:

Blue cheese was discovered entirely by chance by a shepherd tending to her sheep in the alpine pastures of the Pyrenees Mountains, one fateful day a great many years ago.

This young shepherd had taken up into the alpine pastures a cheese sandwich: a piece of fresh sheep's milk cheese layered between two slices of rye bread. But on this important day in history there was a bit of rain in the air, and the shepherd decided to protect her sandwich from the rain by leaving it in a nearby cave. Fate, however, had other plans for her sandwich, for the rain intensified into quite a storm, and in hurriedly getting her sheep to protection in the valley below, the shepherd left the pasture that day without having eaten her sandwich.

It wasn't until many months later that the shepherd was up in that same alpine pasture tending her sheep. And when she passed by the cave she remembered she had left her sandwich there. So she went to investigate and see what had happened to it, and sure enough, the sandwich was still there. But something strange had consumed it.

The two slices of bread had become encrusted with a thick growth of mold and were entirely inedible, but the cheese within was struck through with electric-blue-green veins and was giving off the most intoxicating aroma . . .

According to legend, she was either really curious, or really hungry, for the shepherd tasted the cheese and found it to be delicious. She realized that it was the moldy bread that had given the cheese such interesting qualities, and she took her discovery back to the nearby village of Roquefort, where she instructed cheesemakers there in how to make their cheeses blue.

Though this story may be apocryphal, it sheds light on practices that traditional cheesemakers use to make their cheeses blue: AOC (the French version of PDO) regulations controlling the production of Roquefort require cheesemakers to source their *Penicillium roqueforti* from rye bread left to go moldy in their cheese caves. Even packaged lab-raised *Penicillium roqueforti* fungus is raised on breadlike media, then sold to unsuspecting cheesemakers, who could have just used a piece of moldy bread to start their blue cheeses!

Piercing a cheese encourages the development of blue veins by allowing air—necessary for fungal growth—to penetrate to the interior.

A slice of sourdough bread made blue by inoculating it with *Penicillium roqueforti* from a ripe blue cheese.

- RECIPE -
CULTIVATING PENICILLIUM ROQUEFORTI

Penicillium roqueforti isn't just to be found infecting cheese. One of the more common household fungi, it's an opportunistic species that also grows on decaying fruit, old bread, and leftovers forgotten in the fridge. As it turns these foods "moldy" it produces and disperses its fungal spores, which lie in wait for another opportunity to infect its favorite foods, starting the cycle of decay and spore dispersal again. And if it so happens that some of those wild spores land on a cheese, then we're in for a treat!

One of the most insidious and infectious fungi, wild spores of *Penicillium roqueforti* can ruin entire batches of commercially prepared white-rinded cheeses. Many cheesemakers are troubled by this fungus, installing HEPA filters to remove any *P. roqueforti* spores that may be floating around their dairies and could contaminate their cheeses. Sometimes they even build separate cheesemaking facilities to handle *P. roqueforti*–ripened cheeses, if they make blue cheeses at all. I, however, prefer to embrace this wild fungus and appreciate the little spots of blue that sometimes appear on my Crottins and Camemberts like manna from heaven.

If you want your cheeses to be entirely covered with blue, however, don't just rely on wild *P. roqueforti* spores: Grow your own fungus on a piece of moldy bread!

Of course, not just any piece of moldy bread can be used as a source of *Penicillium roqueforti* fungus, one of many fungal cultures that flourish on bread. A piece of commercial yeasted bread, left in a plastic bag in your cupboard, will play host to dozens of species of wild fungus looking for food to feed on, many of which might contaminate a blue cheese.

Sourdough, though, which is naturally resistant to fungal growth because of its acidic nature, limits the growth of most fungal species; but one of the few species that grows well upon sourdough bread is . . . you guessed it, *Penicillium roqueforti*.

By creating the right conditions, you can grow a pure culture of *Penicillium roqueforti* on a piece of sourdough bread. And whether you are making Roquefort, Stilton, or a surface-ripened blue, you can use your own homegrown *P. roqueforti* spores for all of your blue cheese needs. Here's the method for growing a culture of *P. roqueforti* at home:

INGREDIENTS
1 slice fresh sourdough bread—wheat or rye
1 pea-sized piece ripe blue cheese

EQUIPMENT
Airtight container

TIME FRAME
2 weeks

TECHNIQUE
Get some good sourdough bread. Many varieties of sourdough bread available in the grocery store are made with commercial yeast, so they won't be acidic enough to limit the growth of other fungi. The best way to be sure that the sourdough bread is good? Make it yourself! (The technique for starting a sourdough starter, as well as a recipe for baking good sourdough bread, can be found in appendix A.)

Inoculate the bread with *Penicillium roqueforti*: Take a slice of bread and spread on it a pea-sized piece of ripe blue cheese, preferably a piece marked with blue, the color of the spores of *P. roqueforti*.

Incubate the inoculated bread: Place the *Penicillium roqueforti*–infected bread into a Tupperware container—the perfect container for growing fungus! Seal the container, and let it sit at room temperature for 1 to 2 weeks. Check up on the bread every few days to observe the spread of the fungus: It will start to grow as slightly raised bumps of white. As the fungus advances over the bread, it forms a ring of fresh white mycelial growth surrounding a zone of mature, greenish-blue fungal spore production; this blue/white color combination is an identifying trait of the *P. roqueforti* fungus. Much like when you're aging a cheese, I recommend flipping the bread regularly, and wiping off any excess moisture from the container to ensure that the bread does not get too wet.

Dry the blue bread: Once the piece of sourdough bread is completely enveloped in greenish-blue fungal growth, the "blue bread" is ready. The container can then be opened, and the blue bread left to dry for a few days, flipping it to be sure that the bread is fully dried. Once dried, the blue bread can be kept in a jar and will be preserved for months to years without refrigeration—the fungal spores on it are remarkably stable.

Blue bread showing its blue-green fungal spores.

Use your *Penicillium roqueforti* **spores to make blue cheese:** Break off a pea-sized piece of the moldy bread for every gallon of milk; place that small piece of bread into ¼ cup (60 mL) or so of water; mix the blue bread into the water to release the spores into it; then pour the water through a strainer over the milk at the beginning of the cheesemaking process to introduce the *Penicillium roqueforti* fungal spores into the cheese.

To add blue fungus to a cheese, mix a piece of blue bread in water to release its spores, then strain the water through a sieve into the milk.

To grow *P. roqueforti*, inoculate a piece of sourdough bread with a small bit of blue cheese; leave the bread to incubate in a humid environment; and wait for the fungus to consume the bread before drying it.

- RECIPE -
SURFACE-RIPENED BLUE CHEESE

The simplest and quickest of blue cheeses to make are of the small, surface-ripened variety. Made in much the same way as a Camembert, but inoculated with *P. roqueforti* fungus, surface-ripened blues have the same shape as Camemberts and a similar molten interior but with a bold blue color and surprisingly mild blue flavor.

Because surface-ripened blue cheeses have a high moisture content, large surface area, and small size, they ripen more quickly than other blue cheeses; as the fungus consumes the moist cheeses from the exterior, their flattened shapes help them to ripen fast. Whereas most blue cheeses take several months to develop their characteristic flavor, surface-ripened blue cheeses can ripen in as little as 4 weeks.

This is a great first blue cheese to make. Because of its quick ripening time and small size, it is less intimidating to a novice blue cheesemaker; because of its quick aging and small size, it is also a relatively mild blue cheese, and is less intimidating to a novice blue cheese eater. It's especially interesting and gratifying to ripen cheeses with homegrown *Penicillium roqueforti* culture: Considering what this fungus does to other foods that it infects, its effect on cheese can truly be described as magic!

Ingredients

1 gallon (4 L) good milk
¼ cup (60 mL) active kefir or active whey
Pea-sized piece of blue bread (recipe on page 197)
Regular dose rennet (I use ¼ WalcoRen calf rennet tablet)
1 tablespoon (15 mL) good salt

Equipment

1-gallon (4-L) cheesemaking pot
Knife
Wooden spoon
3 Camembert-sized forms
Draining rack
Cheese cave at 50°F (10°C)

Time Frame

4 hours to make; 1–2 months to age

Yield

Makes 3 small surface-ripened blues

Technique

Warm the milk to 90°F (32°C)—baby-bottle-warm.

Add to the milk the active kefir or whey.

Add *Penicillium roqueforti* fungal spores by mixing the small piece of blue bread in ¼ cup (60 mL) water, then pour the blue water through a sieve over the pot of milk. Mix the cultures in gently.

Incubate 1 hour: Keep the pot of milk in a warm place to encourage the starter bacteria to ripen the milk. Wrap the pot with towels to hold in its heat.

Add a regular dose of rennet, dissolved in ¼ cup (60 mL) water. Mix the rennet in gently.

Incubate 1 hour, keeping the pot at 90°F, to ensure good curd formation.

Cut the curds: Check for a clean break, then cut the curds into ¾-inch (2-cm) cubes in three series of cuts.

Slowly stir the curds to keep them from sticking to one another, and to encourage them to give off their whey. Give the pot a thorough stirring every 5 minutes for half an hour to an hour, until the curds have a firm consistency.

Whey off: Let the curds settle to the pot for several minutes, then pour off the whey.

Form the curds: Ladle the curds, by hand, into the cheese forms, and place the filled forms atop a draining table. Leave the cheeses to firm up for 24 hours, flipping them after 1 or 2 hours of draining.

Salt the cheeses: Take the cheeses out of their forms, and apply 1 teaspoon (5 mL) of salt to the surfaces of each, or place in a salting brine for 1 hour apiece.

To make a surface-ripened blue cheese, ladle basic rennet curds inoculated with *Penicillium roqueforti* into forms; salt and air-dry the cheese; then age it without intervention to encourage fungal growth.

Air-dry the cheeses at room temperature for 1 to 2 days, flipping them twice daily.

Place the cheeses in a cheese cave to age. Blue cheeses prefer a cool aging environment, with temperatures of 50°F (10°C) or less and high humidity.

Check on the cheeses twice a week. Flip them so they don't stick to their aging surfaces, and check to see that they aren't too wet. Fungal growth should start to appear in 1 to 2 weeks. Much like a Camembert, they can be wrapped after several weeks and finished in the cooler environment of a refrigerator. Your surface-ripened blue cheeses will be ripe after 1–2 months of aging.

- RECIPE -
BLUE DREAM CHEESE

If you wish to avoid the use of rennet in cheesemaking, this recipe allows you to have all the fun of aging moldy cheeses without resorting to rennet. Before I learned how to use rennet, Dream Cheese served as my base cheese for aging blue and white-rinded cheeses.

To make Blue Dream Cheese, inoculate yogurt with *Penicillium roqueforti*, then hang it into a Dream Cheese; salt, then air-dry the cheese; then pierce it to encourage blue veins.

This small and tender blue cheese begins as a fresh, firm yogurt cheese. The fresh cheese is intentionally inoculated with spores from *Penicillium roqueforti*. To be sure that the cheese develops blue veins, it's skewered to allow air to pass to the interior, which tricks the fungus into producing its blue spores there. The cheese is then tended to in a cheese cave as it ages for 2 months.

INGREDIENTS

1 quart (1 L) yogurt
Pea-sized piece of blue bread
1 teaspoon (5 mL) salt

EQUIPMENT

Basic cheesemaking equipment
Cheesecloth—preferably a du-rag
Cheese cave at 50°F (10°C) and 90% humidity
Skewer

TIME FRAME

2 days to make a firm Dream Cheese; 1–2 months to age it into a blue cheese

YIELD

Makes a 7 ounce (200 g) cheese

TECHNIQUE

Infect the yogurt with *Penicillium roqueforti*: Prepare a *P. roqueforti* inoculant by dissolving a pea-sized piece of ripe blue cheese or blue bread (with its blue-green spores) in ¼ cup (60 mL) of water. Then pour the water into the yogurt, and mix it in lightly.

Make a firm, round Dream Cheese. Hang the blue-infected yogurt in a du-rag for 24 hours. Keep the cheese in its round shape by regularly tightening the du-rag's tails.

Salt the cheese by opening up the du-rag, mixing the salt into the curd, then tying the cloth back up and tighten it with its ties to press the curd back into shape. Hang the cheese after salting for another 1 to 2 days to dry, tightening it a couple of times with the ties as it hangs.

Age in a cheese cave: Once the cheese is dry to the touch, place your blue-infected Dream Cheese in a cheese cave to age. Check up on it regularly to be sure that it's happy. Flip it every other day so that it's well aerated on all sides.

Skewer your Dream Cheese: After 1 week of aging pierce the cheese half a dozen times with a skewer, ensuring that the fungus gets the air it needs to develop and give your cheese blue veins.

Continue aging the cheese for several more weeks, flipping it twice a week. Watch as the *Penicillium roqueforti* fungus spreads over the cheese and begins to consume it.

Your Blue Dream Cheese will be ready to eat in 1 to 2 months.

- RECIPE -
GORGONZOLA-STYLE BLUE CHEESE

Semi-firm blue cheeses such as Gorgonzola are made in nearly the same manner as surface-ripened blues. They, too, are started with basic rennet curds, and much of their handling is similar. The main differences between the two are that semi-firm blue cheeses are bigger, the curds are salted before forming to ensure plenty of cracks and crevices within the cheese, and the cheeses are skewered to help the *Penicillium roqueforti* fungus ripen them from within and develop their characteristic blue veins.

Though cheesemakers encourage the growth of *P. roqueforti* on the interior of Gorgonzola-style cheeses, they discourage its growth on their rinds. Not as sightly as the delicate blue veins on the interior, blue growth on the rinds can be restricted through several means: Cheesemakers can wash the rinds of their blue cheeses to discourage *P. roqueforti* and encourage raw milk's *Geotrichum* or *Brevibacterium linens* to give the cheeses white or orange rinds; they can wrap their cheeses in foil or a

To make Gorgonozola, start with basic rennet curds; drain them and salt them; then pack them into the form.

clothbinding like cheddar's to stop air from accessing the rinds; or they can trim their cheeses of surface growth by physically removing the rind. Many industrial cheesemakers apply a "natural" fungicide known as natamycin to blue cheese rinds to keep them free from any surface growth (see chapter 20: Gouda for more info on natamycin).

Gorgonzola-style cheeses take longer to age than surface-ripened blues, in part because they have a smaller surface area for their size. As well, the method of salting the curds before forming helps draw out more whey, making a firmer, drier curd that takes longer to age; it is the longer aging time that helps these big blues develop their characteristic flavors.

INGREDIENTS

5 gallons (20 L) good milk
1 cup (240 mL) active kefir or active whey
Hazelnut-sized piece of blue bread with *Penicillium roqueforti* fungal spores
Regular dose of rennet (I use 1 full tablet WalcoRen calf rennet)
Good salt

EQUIPMENT

5-gallon (20-L) cheesemaking pot
Basic cheesemaking tools
Very large colander
Large bowl

Flip the cheese several times in its form until it has knit together; wash the cheese's rind with whey to keep blue fungal growth at bay; then pierce the cheese to encourage fungal growth on its interior.

Draining table
Cheesecloth
1 gallon-sized (4 Litre-sized) cheese form
Cheese cave at 50°F (10°C) and 90% humidity

TIME FRAME
2–3 months

YIELD
Makes a 5-pound (2-kg) wheel of Gorgonzola cheese

TECHNIQUE
Warm the milk to 90°F (32°C)—baby-bottle-warm.

Add the active whey or kefir to the milk.

Add the *Penicillium roqueforti* fungal spores dissolved in ¼ cup (60 mL) water. Stir the cultures in.

Incubate 1 hour. Keep the pot warm, encouraging the bacterial cultures to thrive.

Add rennet: Dissolve a regular dose of rennet in 1 cup (240 mL) water. Mix it into the milk gently.

Incubate 1 hour. Keep the pot of milk warm and undisturbed to ensure good curd development.

Cut the curds: Check for a clean break, then cut the curds into ¾-inch (2-cm) cubes, in three series of cuts. Wait several minutes between each series of cuts to help the curds heal.

Gently stir the curds, every 5 minutes, for 30 to 60 minutes—until they develop a firmness reminiscent of poached egg whites. Keep the temperature at 90°F.

Allow the curds to settle for several minutes, then whey off.

Drain and salt the curds: Scoop up the curds into the large colander set over a bowl. Sprinkle ½ cup (125 mL) of salt over the curds and lightly stir it in with your hands. Let the curds drain for 5 minutes, stirring them regularly to prevent them from knitting together.

Form the curds: Scoop the curds from the draining table into a cheesecloth-lined cheese form. After 10 minutes, carefully flip the cheese so that it forms evenly on each side. Leave the curds to form for 24 hours, flipping the cheese occasionally to keep its shape even, remove the cheesecloth once the cheese holds together.

Prepare a light brine: Save 1 quart (1 L) of whey, and add 1 tablespoon (15 mL) of salt to it to make a washing brine. Keep this in the cave for washing the rind of the cheese.

Air-dry the cheese: Remove the cheese from its form, and leave it to air-dry for 24 hours on a draining rack flipping it once or twice.

Place the cheese into a cave to age. Ideal conditions are 50°F (10°C) or less and 90 percent humidity.

Age the cheese, washing its rind: For the first week of aging, rub down the rinds of the blue cheese with a light salt brine every other day to discourage the growth of *P. roqueforti* if you choose. Flip the cheese every time you wash.

Pierce the cheese: Once the cheese has aged for a week, it can be pierced. With the skewer, pierce the cheese several dozen times, taking care to evenly distribute the piercings. Skewer the cheese from both the top and the bottom as well as the sides.

Continue aging without washing: Place the cheese back in its cave and continue to age it, flipping it twice weekly to prevent it from sticking. Your Gorgonzola should be ready in 2 to 3 months.

CHAPTER EIGHTEEN

Washed-Rind Cheeses

Fungus will grow on all cheeses, even if it's not intentionally introduced. Indigenous fungi from raw milk and the fungal flora of the cheesemaking environment will all collaborate to transform a freshly made cheese into a fungally ripened one. Even a cheese made from pasteurized milk, prepared in a sterile manner with DVI cultures and aged in a sanitized cave, will come to grow a fungal coat as a result of inevitable wild fungal "contamination."

Washed-rind cheeses are a class of cheeses whose rinds are washed to control unwanted fungal growth. When a cheese is washed, however, not only is fungus kept in check, but a new regime of cultures thrives in the open ecology left by the washing.

As the fungal growth is set back, the wet, salty skins of washed-rind cheeses support the growth of a complex ecology of bacteria and yeasts that comes to dominate their development. The most celebrated culture of this washed-rind ecology is a bacterium known as *Brevibacterium linens*.

Brevibacterium linens

Brevibacterium linens stinks! Considered the most noxious ripening culture, *B. linens* ripens the smelliest of cheeses: Limburger, Muenster, and Epoisses are all made with a regular rind washing that encourages the growth of putrescent *B. linens* bacteria. But as famous as they are for their unmistakably footlike aromas, washed-rind cheeses can have surprisingly agreeable flavors.

Brevibacterium linens ripens cheeses from the rinds, much as *Penicillium roqueforti* and *Geotrichum candidum* do. An aerobic surface-ripening culture, *B. linens* thrives on the rinds as it feeds on a cheese. And as it feeds, the cheese is slowly transformed, taking on the qualities that washed-rind cheeses are known for.

B. linens–ripened cheeses are also recognized for their beautifully colored rinds. The many different colors of washed-rind cheeses range from pink to orange to burgundy. No added coloring is needed to develop the brightness of these cheeses: *B. linens* adorns them with its own colorful pigments.

B. linens is a common bacterial culture that thrives in the moist and salty conditions created by a regular washing. But it doesn't just grow on washed-rind cheese; it also thrives in other moist and salty ecologies . . . such as those found between your toes! These cheeses are known for their footlike aromas precisely because the *B. linens* culture that is responsible for their aroma is one of the cultures that give our feet their aroma. And though *B. linens* is discouraged from growing on our feet through regular washing, cheesemakers regularly wash their cheeses to encourage the growth of the very same culture.

The Washing Process

To wash a cheese, cheesemakers regularly rub their ripening rinds with a cloth dipped in lightly salted

Smearing the rind of a ripe washed-rind cheese inoculates the cloth with *B. linens* cultures; the cloth then serves as a source of *B. linens* culture for newly made washed-rind cheeses.

brine. The brine is prepared from fresh whey, or some other tasty liquid, and a small amount of salt (see the recipes, page 214).

Also known as smear-ripened cheeses, washed-rind cheeses are thoroughly smeared to physically set back any established fungal growth. Cheesemakers scrub their washed-rind cheeses very firmly!

All sides of a cheese are smeared to limit fungal growth. If any part is left unwashed, fungal ecologies will come to dominate that part: Any indents spared the smear will develop fungal ecologies, often showing the white colors of beneficial fungi that thrive alongside *B. linens.*

A washing every other day early in a cheese's development sets back fungus on the rinds and creates the conditions needed for the growth of *B. linens* cultures. After about one month of regular washing, fungal growth is overcome, a bright orange *B. linens* culture establishes itself on the rind, and the cheese need no longer be washed.

Cultivating *B. linens* Cultures

Much like all the other cultures they use, contemporary cheesemakers inoculate their washing brines with freeze-dried *B. linens* bacteria to ensure the appropriate microbiological development on the rinds of their cheeses. Traditional cheesemakers, of course, have another way of doing things.

Keeping B. linens *at Home*

Traditional cheesemakers keep their washed-rind cultures through their culture of washing their cheeses' rinds. The methods they use cultivate the microorganisms that define the washed-rind ecology.

Already ripened washed-rind cheeses are an excellent source of *B. linens* cultures, and cheesemakers need only wash the rinds of a mature cheese to inoculate their washing cloths with the appropriate cultures. Cheesemakers use the same cloth to smear the rinds of all of the cheeses they wash, and thus they carry the *B. linens* cultures to every cheese they smear.

B. linens cultures can thus be kept on the rinds of ripening washed-rind cheeses in a cheese cave.

Beginning cheesemakers who wish to make washed-rind cheeses only need a small piece of ripe washed-rind cheese as a source of culture to use as an inoculant (just ask your cheesemonger for a sample!). And once they've established *B. linens* ecologies on their own cheeses, they can use their own cheeses as inoculants.

However, even if you don't wipe the rinds of a mature washed-rind cheese first, you'll still find that your cheeses develop *B. linens*–influenced rinds as a result of the wild populations of *linens* on your skin, in raw milk, and in unprocessed sea salt. A nonsterile cheesemaking helps to cultivate the appropriate microorganisms on the rinds of washed-rind cheeses.

Diversity in Washed-Rind Cheeses

It is a bit misleading to suggest that *Brevibacterium linens* is the only culture responsible for the development of washed-rind cheeses. No one culture can take the credit for the transformative effects of rind washing. A community of microorganisms, including diverse species of bacteria, fungi, and yeasts, work together to make washed-rind cheeses bloom. You may notice signs of fungus beginning to blossom on the rinds of washed-rind cheeses if the rind washing is stopped. The culture that you are witnessing is *Geotrichum candidum,* one of many fungal, bacterial, and yeast cultures that live in balance on the surfaces of washed-rind cheeses.

Brevibacterium linens itself may not even be present on the rind of a washed-rind cheese. Many other bacterial species grow in washed-rind ecologies and produce similar colors and flavors.

DVI Brevibacterium linens

Brevibacterium linens is yet another packaged freeze-dried culture that commercial cheesemakers purchase to make their cheeses. In using freeze-dried *B. linens* culture, though, cheesemakers add only the one species of *B. linens* to their washed-rind cheeses, not the diverse profile of cultures needed to establish healthy rind ecologies.

In relying on monocultural freeze-dried *B. linens* culture to make their washed-rind cheeses, cheesemakers leave ecological voids in their cheeses that make them susceptible to contamination. As a result, many species of wild bacterial and fungal cultures not intentionally added by cheesemakers establish themselves on the rinds of commercially prepared washed-rind cheeses.

Present-day cheesemakers are discouraged from practicing traditional methods of cultivating *B. linens* cultures because of the fear of transferring foodborne illness from one batch of cheese to another. Fear-based recipes for making washed-rind cheeses call for inoculating the washing brine with freeze-dried *B. linens* culture, then washing each batch of cheese with a sterilized cloth dipped in the brine, keeping each batch of cheese in isolation from all others.

The restrictions on such traditional cheesemaking practices may be warranted in the industrial production of cheeses, but not on the small scale. In large dairies, milk from hundreds of cows may be mixed together, increasing the possibility that a batch of cheese could carry disease; if cheesemakers use the same cheesecloths to smear several batches of cheese, the risk of cross-contamination grows dramatically as every cheese is washed. However, with small-batch cheesemaking, this concern of contamination is significantly lessened.

Variations on Washed-Rind Cheeses

Though they make up a class of cheeses all their own, washed-rind cheeses can be made of any basic curd style: lactic curd, soft curd, or Alpine curd; even cheddars and Goudas can have their rinds washed as they age. And though they could all be considered washed-rind cheeses, cheeses from these classes become vastly different from one another when washed. But generally, only smaller and softer smeared cheeses are classified as washed-rind cheeses because they are the most sensitive to the effects of the washing process.

The moisture content, the surface-area-to-volume ratio, and the acidity of the curd all have significant effects on the development of washed-rind cheeses. A hard Alpine cheese, when washed, will ripen quite differently from a soft rennet curd cheese; and a small washed-rind cheese will ripen differently from a large washed-rind cheese.

Small washed-rind cheeses with much surface area and made of soft rennet curds develop molten interiors and strong ammonia flavors much like a Camembert. The *B. linens* cultures affect the ripening cheese in a similar way to *Geotrichum* and *P. roqueforti*: The bacteria consume the lactic acid in the cheese and make the curd less acidic. As the curd becomes more alkaline, the calcium comes out of it and the flesh of the cheese begins to liquefy.

A large Alpine cheese, however, will ripen differently when washed. A regular washing will result in the development of a bright-orange *B. linens* rind; however, the low moisture and large size of the cheese keep the *B. linens* culture from softening up its flesh. A washed-rind Alpine cheese will have the same firm texture as any other Alpine cheese. Even its aroma and taste will only be mildly influenced by the stinky *B. linens*.

The way that the brine is prepared can also have a strong influence in the development of washed-rind cheeses.

Recipes for Brines

The brines used to wash cheeses' rinds are usually prepared of fresh whey and salt. To make a basic brine, dissolve 1 tablespoon (15 mL) of salt in 1 quart (1 L) of fresh whey. The saltiness of this brine roughly equals the saltiness of the cheeses being washed.

Brines can also be prepared with a higher salt concentration; however, this will change the effect of the washing. If you add ¼ cup (60 mL) of salt to a quart (1 L) of whey, the saltier washing brine will cause the cheese to age differently. A less salty brine will result in a pinkish rind, while a saltier brine will produce a more vivid orange rind. Saltier brines also seem to make washed-rind cheeses less stinky. The different salt concentrations influence

the development of the bacterial and fungal cultures on the rinds, which in turn affects the colors that develop on the cheeses' surface.

Brines can also be made with tasty liquids, such as beer, wine, or spirits. You can use your favorite alcoholic beverage to wash the rinds of your cheeses to create the perfect partner for your favorite drink. You can even wash the rinds of your cheeses with your home-brewed beers or wines and diversify the scope of your home-brewing activities.

To prepare a brine with a mildly alcoholic beverage, simply add 1 tablespoon (15 mL) of salt to 1 quart (1 L) of the drink. If you are using spirits such as brandy or whiskey, dilute the spirits 4:1 with whey to reduce the alcohol content of the brine, which could negatively affect beneficial bacterial growth. Sulfites in wine or cider that are added to control the growth of yeasts may also inhibit the growth of beneficial microorganisms on washed-rind cheeses.

Generally, only whey or alcoholic beverages are used to wash cheeses: Sweet fruit juices contribute too much sugar and can cause ripening problems.

Water is not usually used in washing brines, because it leaches minerals from a ripening cheese.

White Rinds and Rind Washing

Rind washing is also the traditional way to cultivate white-rinded cheeses such as Brie and Camembert. The rind washing eliminates the growth of *Penicillium roqueforti* and other colorful fungi and creates conditions conducive to the growth of white *Geotrichum candidum*.

If the rind washing is stopped after only a week of aging, the yeasts and fungi that are also encouraged by it can come to dominate the development of the cheese in place of the *B. linens* bacteria, resulting in a pure white *Geotrichum* rind.

This method of cultivating white rinds is more effective and more natural than the standard approach, which calls for sterilization, then inoculation with freeze-dried *Penicillium candidum*. For more on washing white-rinded cheeses, see chapter 16.

– RECIPE – LIMBURGER

This soft, square-shaped, pinkish German cheese is famous for its funk. Its strong flavor makes it an excellent partner for dark German rye breads and dark German beer.

Limburger develops the strongest aromas of all washed-rind cheeses because of its small size and large surface area: its flattened square shape encourages surface ripening, which, from the prevalence of *B. linens* on its rind, results in an extra-strong stench. Using a lightly salted brine helps Limburger develop a pinkish color and strong aroma.

Kefir can be used as a starter culture to get this square cheese rolling. Containing a diverse population of beneficial bacteria, yeasts, and fungi, kefir also serves as an inoculant for many of the cultures that help to develop a washed-rind cheese. Wiping the rinds of already ripened washed-rind cheeses before wiping the rinds of the newly made Limburger can help to further reinforce the development of the *B. linens* ecology.

This recipe for Limburger can be adjusted to make other washed-rind cheeses. If you wish to make an Epoisses-style cheese, for example, shape the curds in small, round Camembert forms, and wash the rinds with Marc de Bourgogne (a spirit distilled from Burgundy wine) diluted 4:1 with whey. You can also make your own washed-rind cheeses that are truly products of your land by washing the cheeses with tasty drinks from your terroir. My favorite washing liquids are the brews I make at home—wild fermented apple ciders and blackberry wines.

INGREDIENTS

1 gallon (4 L) good milk
¼ cup (60 mL) active kefir or whey
Regular dose rennet (I use ¼ tablet WalcoRen calf rennet for 1 gallon milk)

To make Limburger, start with a square-shaped rennet cheese; inoculate a cloth with *Brevibacterium linens* culture from a ripe washed-rind cheese; then wash the rinds of the young Limburger as it ages to encourage the growth of the appropriate cultures.

Salt
Already ripe washed-rind cheese to use as
   an inoculant
1 quart (1 L) washing brine

EQUIPMENT

1-gallon (4-L) stainless pot
Wooden spoon
Long-bladed knife
3 cheese forms, preferably 4 inches (10 cm) square
Draining rack setup
Cheese cave at 50°F (10°C) or less and 90% humidity
Cheesecloth for smearing rinds

TIME FRAME

1–2 months

YIELD

Makes 3 Limburger cheeses

TECHNIQUE

Slowly warm the milk to 90°F (32°C), baby-bottle-warm.

Add the active kefir or whey, and mix into the milk lightly.

Incubate 1 hour, covering the pot and keeping the milk at 90°F to encourage the bacterial cultures to ripen the milk.

Add the rennet by dissolving the regular dose in ¼ cup (60 mL) water and gently mixing it into the milk.

Incubate 1 hour, covering the pot and keeping the milk at 90°F to encourage the rennet to set the curd.

Check the curd for a clean break. Proceed to cut the curd once it has set.

Cut the curd, in three series of cuts, into ¾-inch (2-cm) pieces. Wait a few minutes between each series of cuts to let the curds heal.

Keep the pot at 90°F while you gently stir the curds for 30 to 60 minutes, to encourage them to firm. Proceed to whey off once the curds have the firmness of a well-poached egg.

Whey off: Pour the whey off the pot, reserving a quart for making a washing brine.

Strain the curds, by hand, into their forms. Fill the forms right up to the top.

Prepare a light salt brine for washing the rinds by mixing 1 tablespoon (15 mL) of salt into 1 quart (1 L) of the still-warm whey. Pour this into a sealable container, and keep it in the cave to use for future rind washings. Alternatively, you can use beer, wine, or cider in place of whey.

Place the filled forms on a draining rack, and let them drain for 24 hours. Flip the cheeses after 2 hours to give them an even shape. Remove the cheeses from their forms once they're firm.

Salt the cheeses: Apply around 1 teaspoon (5 mL) of salt to the surfaces of each cheese. Be sure to spread the salt around the entire cheese, and apply more if some falls off.

Air-dry the cheeses, and let them drain for at least 24 hours. Flip them once or twice as they dry.

Place the cheeses in the cave to age when they are dry to the touch.

Wash the ripening cheeses' rinds twice weekly for 1 month: Smear the rind of a mature washed-rind cheese with the cloth moistened in the washing brine before washing the rinds of the young cheeses to inoculate them with *B. linens* and its associated cultures. Flip the cheeses when you put them back in the cave to age.

Let the cheeses continue ripening for several more weeks. Once you see signs of the *B. linens* culture establishing itself on the cheese, you can stop the rind washing. At this point the *B. linens* culture will continue to thrive, but its partner fungus, *Geotrichum candidum*, will begin to bloom on the surface. Continue to flip the cheeses twice a week as they ripen so that they do not stick to their ripening surfaces.

Limburger cheese is ready to eat when it has developed its stink and bright color, typically after 4 to 8 weeks of ripening.

CHAPTER NINETEEN

Alpine Cheeses

The Alpine regions of Europe are linked by a common cheesemaking tradition. In the French, Italian, and Swiss Alps, milking animals are herded into the mountains to graze on the high pastures, and cheesemakers spend the summer alongside the animals, transforming their milk into giant wheels of Alpine cheese.

The Tradition of Alpine Cheesemaking

Alpine cheeses arose from the seasonal trans-humance cultures of these mountainous regions. In the summer months, farmers leave their milking animals in the care of herders, who lead their charges up into the alpine meadows. The cows feast on the lush mountain grasses, and the herders tend to the animals, take their milk, and transform it into cheese. And because they're a long distance from the markets in the valleys below, the herders follow the Alpine cheesemaking method to preserve as much milk as possible into huge firm cheeses that are long lasting and easily transportable.

The massive cheeses are aged in the cellars of the cheesemakers' alpine huts, and by the end of the grazing season their cellars are filled with a summer's worth of cheeses. In the fall the herders descend from the mountains with the animals and their cheeses, return them to their farmer, and, historically, reserve a few cheeses as payment for their work.

Nearly every valley in these regions is known for its distinct cheeses that arise from the unique soil chemistry and unique plants that the animals feed on, the different breeds of animal that make that cheese, and the particular way that each valley's cheeses are handled and aged. Far more than any other cheese, Alpine cheeses are the purest expression of the flavor of their milk, and the perfect picture of their distinct terroir.

Raw Milk and Alpine Cheeses

Alpine cheeses are the only class of cheeses that are still consistently made with raw milk. Those who make Alpine cheeses know that raw milk results in the best expression of taste and place in their cheese. The essence of the milk is most concentrated in Alpine cheeses, and as the cheese is aged, the milk's flavor blossoms.

The milk used to make Alpine cheeses must therefore be of the highest quality. Generally, cheese-makers only make Alpine cheeses when their animals are feeding on green pasture in the summer months. If their animals are feeding on grain, hay-lage, and silage, as is the case with confined dairy cattle, the cheesemaking quality of the milk is greatly reduced, the curds do not stand up to the rigors of the Alpine cheese method, and the flavor of the cheese suffers.

Pasteurization, too, weakens the strength of the milk and results in a curd that does not perform as it should. Heat treatments also devastate the diverse flora of raw milk, along with the milk's endemic enzymes that together develop Alpine cheeses' desired character.

Alpine cheeses can be made with cows' milk, goats' milk, or sheep's milk. Each milk will respond similarly to the Alpine cheesemaking method; however, the cheese from each species will be dramatically different from the others' because of the different flavors, enzymes, and microbiological communities in each animal's milk. Using raw milk helps ensure that distinctiveness.

Considerations for Making Alpine Cheeses

The Alpine cheese method is one of several techniques for making firm cheeses. Two other methods, cheddaring and washing curds with hot water, will be explored in later chapters.

The Alpine cheese method is an adaptation of the basic rennet cheesemaking method. Cheesemakers culture their milk, set the curd with rennet, cut the curd, stir the curd, then form the curd, much as with softer rennet cheeses. But what distinguishes Alpine cheeses is that they are made with many gallons (many, many litres) of milk and coagulated with extra rennet; their curds are cut very small and cooked in large cauldrons. The firm curds that result are then pressed into enormous wheels of cheese.

Because of these distinctions, certain considerations must be taken when making Alpine cheeses.

Amount of Milk

Alpine cheeses need a lot of milk. Made massive to preserve as much milk for as long as possible, Alpine cheeses often weigh in at over 50 pounds (20 kg) apiece—more than 60 gallons (240 L) of milk is needed to make an Alpine cheese that large.

The larger the cheese is, the firmer it will be, the less moisture it will lose as it ripens, and the longer it will age. And the longer an Alpine cheese ages, the more interesting flavors it develops: The very tastiest Alpine cheeses are also the largest! Two

gallons (8 L) of milk is the absolute minimum amount to make one Alpine cheese, though for best results I recommend at least 5 (20 L). Alpine cheeses that are too small will dry out before they have a chance to develop their finest possible flavors.

Culture Used

Contemporary cheesemakers often add to their milk strains of thermophilic bacterial starter cultures when they make Alpine cheeses, as these cultures can withstand the higher temperatures to which the curds are cooked. Packaged DVI thermophilic cultures, however, do not offer enough of a range of cultures to help these cheeses develop their characteristic flavors, and cheesemakers must also add DVI ripening cultures. But even if you make Alpine cheeses with pasteurized milk and thermophilic starter and ripening cultures, your cheeses won't take on the complexities of their raw milk counterparts.

Alpine cheeses made with raw milk do not need any added starter culture; the indigenous raw milk cultures provide all the starter and ripening bacteria an Alpine cheese needs. Raw milk, with its native thermophilic bacteria, has cultures that are ideally suited to the making of Alpine cheeses. Raw milk also contains many different species of bacteria, yeasts, and fungi that help Alpine cheeses develop their best flavors as they age. If you wish to make an Alpine cheese without starter culture, allow fresh raw milk to ripen overnight at room temperature before the make to encourage the development of milk's endemic bacterial cultures, which contribute the acidity needed to set the curd. This is a traditional practice that works for this class of cheese; Alpine cheesemakers often leave the evening's milk to sit overnight and mix it with the morning's milk when starting their cheeses.

Kefir, with a similar microbiological profile to raw milk, can also be used as a starter culture to help ensure consistency in the production of Alpine cheeses, as can recycled whey from a previous batch of cheese.

Rennet Dose

Cheesemakers add up to double the recommended rennet dose compared with basic curds when making Alpine cheeses. Very soft curds such as chèvre are made soft with a lighter dose of rennet; firm curds are made firm with a higher dose. The rennet dose develops the curds' strength so that they can resist the vigorous cutting and stirring and the high-temperature cooking that they are subjected to.

Curd Cutting

To make a firm Alpine cheese, the curd is cut to a very small size. Cheesemakers use wire-whisk-like tools to cut their curds to the size of small lentils. The curd is cut quickly, and without remorse, unlike the gentle cutting of basic rennet curds. Because the curd is made firmer with a higher dose of rennet, though, it is not harmed by such handling.

Their small size gives the cut curds much more collective surface area in the pot, which encourages more whey to flow from them. The small curds also firm up faster when cooked, giving Alpine cheeses their dry flesh and hard texture.

Cutting the curd to a very small size defines the making of Alpine cheese.

Cooking

Alpine cheeses are considered "cooked-curd" cheeses because their curds are subjected to a higher-temperature cooking stage. Cheesemakers raise the temperature of the pot to around 110°F (43°C) after cutting, which helps the curds to expel more whey and makes them even firmer.

Alpine cheeses are cultured and coagulated at 90°F (32°C), just like basic rennet curds— the ideal temperature for acid development and curd formation. But once the curds have formed, they are cooked to a higher temperature by slowly heating the pot to 110°F. If the temperature rises too quickly, the curds may stick together in the pot.

Stirring

Cheesemakers vigorously stir their curds as they cook. The stirring ensures that the curds don't stick together and encourages them to give off as much whey as possible.

Your hand is the most effective stirring tool for making alpine cheeses; it has the perfect shape for stirring and can break up clusters of curd if they form. By detecting the clusters early, your hand can tell you if the pot is being heated too quickly. Your hand is also very helpful at gauging the temperature of the pot as the curds cook; if you can no longer keep your hand in the pot, the temperature is over 110°F (43°C), and too hot for the curds as well!

Using your hand to stir also keeps you present, which is extremely important while making this cheese; you do not want to leave the curds unstirred for very long, as they will quickly knit together into a lump of cheese and overcook at the bottom of the pot.

Pressing

The curd, made extra-firm by the extra dose of rennet, the small curd size, and the high-temperature cooking, will not knit together on its own like a softer curd. Pressing is therefore essential when forming Alpine cheeses.

Cheeses can be pressed in a mechanical press, or simply by putting weight atop the cheese in its form: using buckets of warm whey to press your cheeses can achieve as effective a pressing as a mechanical press with much less weight. Warm whey keeps a cheese warmer for longer, increasing the impact of the pressing; once the curds have cooled, the pressing loses its effect.

Cheeses are flipped several times as they are pressed to encourage an even shaping. And the weight of the press is increased every time the cheese is flipped so that the cheese is not squeezed out of its form during the early stages of pressing. If you're using nested plastic containers as a cheese press (my preferred method), fill the press halfway with warm whey for the first pressing, then increase the amount of whey every time you flip the cheese until the container is full.

Salting

Alpine cheeses can be surface-salted or salt-brined much like all other cheeses. Depending on the desired firmness and dryness of curd, some Alpine cheeses are salted more than others.

Alpine cheeses are generally salted immediately after pressing. Salting helps to get out more moisture and stops the development of raw milk bacteria that can generate gas and lead to the formation of "eyes" (think of the holes in Swiss cheese).

To salt an Alpine cheese, apply 1 tablespoon of salt for every gallon's worth of milk used to make the cheese (that's 15 mL of salt for every 4 L of milk); for example, if a cheese was made with 5 gallons of milk, apply 5 tablespoons of salt (75 mL of salt for 20 L of milk). Spread the salt around the entire cheese, and apply more if some falls off. Larger cheeses may need to be surface-salted several times to apply enough salt—only so much can be applied at a time. Alternatively, Alpine cheeses can be brined in a saturated salt brine 4 hours for every pound (that's 8 hours for every kg).

Aging

Alpine cheeses are aged longer than almost every other style of cheese. It takes time for these hard cheeses to develop their characteristic flavors; their low moisture content brings bacterial and fungal development nearly to a halt, and encourages a slow and complex flavor development.

Surface effects are limited by Alpine cheeses' enormous size. And because the cheeses are drier and more mineralized, the surface flora affect the paste of the cheeses less: Alpine cheeses do not liquefy in the same way as smaller and softer cheeses like blues and Camemberts. Alpine cheeses therefore age mostly from the interior. As a result, their development is dominated by the many species of bacteria and numerous enzymes in the cheeses' flesh—which is the reason using raw milk, with its native enzymes and bacteria, is so important to this class of cheeses.

Alpine cheeses can be given a vast array of surface treatments as they age, but their effects are less definitive to the development of a cheese and lean toward the aesthetic, having mostly an effect on the appearance of the rind. Alpine cheeses can be cured at room temperature for several days before aging to grow their natural *Geotrichum* coats; they can have their rinds washed to give them beautiful pink or orange colors (depending on the concentration of salt in the brine); they can be brushed to give them a dry rind; they can be oiled to give them a clean, shiny rind; or they can be given a coat of wax or fat to eliminate surface-ripening effects altogether. The recipe below for Tomme describes how to wash an Alpine cheese's rind to grow a white coat.

Grana Cheeses

Even firmer and more massive than Alpine cheeses are grana cheeses. A variation on the Alpine cheese-making method developed by monastic orders of cheesemakers in the north of Italy, the most *famoso formaggio*, Parmigiano Reggiano, belongs to this class of extra-firm cheeses.

Several factors distinguish grana from Alpine cheeses:

An extra-high-temperature cooking to 135°F (57°C) helps the curds achieve a very firm structure. The firm, dry curd allows grana to be aged for exceptionally long periods.

Grana cheeses are made to be even larger than Alpine cheeses: Their enormous size (up to 100 pounds—that's around 45 kg) helps ensure that they age slowly and ponderously. Rare is the single dairy that can produce enough good milk in a day to fill the several-hundred-gallon vats that are used to make these cheeses; grana cheeses are generally made at cooperative cheesemaking facilities, where farmers pool their milk.

As the cheeses are aged, they are wiped with oil to preserve their pristine rinds. Fungal growth and surface-ripening bacteria are thus kept in check, and the cheeses ripen almost entirely from within.

Grana cheeses' long aging period (often several years) helps them develop their superb flavors and textures. Cheese crystals often grow in grana, crystallizing out of the curd in a slow, almost geological process as the cheeses age.

Parmigiano Reggiano

Parmigiano Reggiano is the granddaddy of grana cheeses.

This superb cheese is the purest expression of milk's potential. Made with the highest-quality raw milk, natural calf rennet, and backslopped with whey as the only starter, Parmigiano may also be the most widely celebrated naturally and traditionally made cheese in the world.

Sadly, the making of Parmigiano Reggiano is beyond the grasp of small-scale cheesemakers. Making this cheese right requires making it more massive than you can imagine. The minimum amount of milk needed for a single Parmigiano is about 125 gallons (500 L), and cheesemakers are required by PDO regulations to make two at a time in gigantic copper vats. So if you don't have a colossal copper cauldron along with 250 gallons (1000 L) of good milk, don't expect your cheese to taste as it should!

If you wish to make a smaller, Parmigiano Reggiano–style cheese at home, here's what to do. Follow the recipe for Tomme, below, making the following adjustments: Use a minimum of 5 gallons (20 L) of milk per cheese; cook the finely cut curds very slowly while stirring vigorously until the temperature

reaches 135°F (57°C); after cooking, let the curds settle in the pot for 30 minutes to cool and coalesce; salt the cheese by brining 5 hours for every pound (10 hours per kg); wipe the cheese with an olive-oil-soaked cloth twice a week as it ages; and age for a minimum of 6 months. A cheese made this way will have many of the same characteristics as Parmigiano Reggiano but won't be able to be aged as long, will not have the same texture, and will lack the depth of flavor because the cheese isn't made quite the same . . . and of course, there's also the issue of the name.

Any cheeses produced in the style of Parmigiano Reggiano but made outside of the PDO region of production in the Po Valley of northern Italy cannot be called Parmigiano Reggiano; even the name Parmesan is protected in the European Union. In North America that is the name you'll have to settle for if you choose to make this cheese.

> **- RECIPE -**
> # TOMME

Tomme (pronounced *tome*) is a small Alpine-style cheese with a natural white fungal coat from the mountainous Savoie region of France.

If you're new to hard cheeses, Tomme is an excellent cheese to try: Its small size makes it less intimidating to the novice cheesemaker; it is relatively simple to make and easy to age; and despite its simplicity, this cheese develops excellent flavors that are sure to impress. This is my go-to hard cheese; when I find myself blessed with many gallons of milk, Tomme is what I make.

I usually start my Tommes with kefir culture because it helps grow a coat of *Geotrichum* fungus and provides diverse bacterial cultures that can help the cheeses develop interesting flavors as they age. Whey saved from a previous batch of Tomme can also be used as a starter; so can native raw milk cultures (let raw milk mature overnight at room temperature before you make the cheese).

A regular rind washing during the first week of a cheese's development controls unwanted fungal growth and encourages the growth of white *Geotrichum candidum*, the preferred fungus for this cheese. This basic Alpine cheese can be ripened in other ways, but each other method will result in a different cheese: You can wash the rind for a longer period to encourage *Brevibacterium linens*; you can oil or dry-rub the rind to limit fungal growth; or you can submerge the cheese in wax to stop any surface-ripening effects. (See chapter 20 for more info on waxing.)

INGREDIENTS

5 gallons (20 L) good cows', goats', or sheep's milk
1 cup (240 mL) active kefir or whey
Double dose rennet (I use 2 tablets WalcoRen calf rennet)
Good salt

EQUIPMENT

5-gallon (20-L) pot
Wooden spoon for stirring
Wire whisk for cutting the curds
Large cheese press follower and form
Draining rack setup
Cheesecloth for pressing
Cheese cave at 40–50°F (4-10°C) and 90% humidity
Cheesecloth for smearing rinds

TIME FRAME

6–12 months

YIELD

Makes a 4-pound (2-kg) Tomme (or two 4-pound Tommes with sheep's milk!)

TECHNIQUE

Slowly warm the milk to 90°F (32°C), baby-bottle-warm.
Add the active kefir or whey, and mix it in lightly.
Incubate 1 hour, covering the pot and keeping the milk at 90°F to encourage the bacterial cultures to ripen the milk.

To make Tomme, cut its curds to the size of a lentil with a wire whisk; stir the curds while cooking them; and let the curds knit together at the bottom of the pot.

Add the rennet: Dissolve the double dose of rennet in 1 cup (240 mL) of water, and gently mix it into the milk.

Incubate 1 hour, again covering the pot and keeping the milk at 90°F to encourage the rennet to set the curd.

Check for clean break. Proceed to cut the curd once it has set.

Cut the curd to the size of a lentil: Using the wire whisk, whisk the curd to within an eighth of an inch of its life (that's 3 mm). A minute or two of brisk whisking will achieve the appropriate curd size: about the size of a small lentil. Stir the curd to be sure that all the curd has been thoroughly cut.

Cook the curds to 110°F (43°C) while stirring. Once the pot has reached 110°F, continue stirring nonstop for 30 to 60 minutes, to encourage them to firm. Once the curds, when pressed between thumb and forefinger, bounce back to their original shape, they are ready to be put into their form.

Pitch the curds. Let the curds settle in the pot for 5 minutes. As they settle, the curds will knit together into a lump of cheese at the bottom of the pot. Do not heat the pot during this time, as the curds may overcook.

Set up your cheese press atop the draining table. Lay cheesecloth in the form to help hold the curds together.

Whey off: Pour the whey off the pot, reserving a quart (1 L) for making a washing brine. Save the rest for ricotta (see the recipe in chapter 22); whey from Alpine cheeses will give the best results!

While they're still warm, strain the curds, by hand, into the form. Fill the form to within an inch of the top, and place the form on the draining rack to press.

Place the follower—half filled with warm whey—atop the cheese to serve as a press. Press the cheese for 5 minutes.

Lift the curd out of the pot and place it into the press; press the cheese with whey; and flip it several times to ensure an even shape.

Prepare a washing brine: Mix in 1 tablespoon (15 mL) of salt into 1 quart (1 L) of the still-warm whey. Pour into a sealable container, and keep in the refrigerator to use for future rind washings.

Flip the cheese: Take it out of the form with the aid of the cheesecloth, and flip it so that it can be pressed on the other side. Rewrap the cheese in its cheesecloth to help ease it back into the form. Add a bit more weight to the pressing.

Press the cheese, flipping it every 10 minutes for about 1 hour. After an hour or so, the cheese will have cooled and stabilized and can be left in its form, unpressed, overnight.

Form the cheese: Leave the cheese in its form without cheesecloth to drain overnight.

Surface-salt the Alpine cheese with 5 tablespoons (75 mL) of salt—1 tablespoon (15 mL) per gallon (4 L) of milk.

Air-dry the cheese for at least 24 hours after salting, flipping it once or twice as it dries. Once the cheese has no visible moisture on its surface, place it into your cheese cave to age.

Wash the ripening cheese's rind every other day for 1 week: Smear the rind with a cheesecloth dipped in the washing brine to encourage the growth of *Geotrichum*. Flip the cheese when you put it back in the cave to age.

Let the cheese continue ripening for several months. Check up on it, and flip it weekly as it ages. Leave the Tomme to ripen for at least 6 months.

CHAPTER TWENTY

Gouda

This famous Dutch *kaas* starts off just like most other rennet cheeses. What defines Gouda is that its curds are washed with hot water to firm them up.

The act of washing the curds with hot water originates from the practice of making of cheese in wooden vats that could not be heated. To warm the vats, cheesemakers would remove some of the whey and replace it with hot water, thus heating the curds and encouraging them to firm. The cooked curds were then pressed into wheels of cheese that could be made larger and aged longer than softer, uncooked cheeses.

Though contemporary Goudas are made in stainless vats that can be easily heated (wooden vats are now, sadly, taboo), Gouda is still made according to the traditional practice of washing the curds with hot water because that's what defines this cheese. Simply heating the curds would not have the same effect: Washing the curds dilutes the cheese's lactose and therefore changes the nature of its evolution. With less lactose, the cheese develops less lactic acid as it ages, and the result is a cheese with a milder, less acidic flavor: Gouda.

Washing the Curds

To make Gouda, cheesemakers first transform their milk into curds according to the basic rennet curd method: The milk is warmed to 90°F (32°C), cultured, and incubated; the cultured milk is renneted

with a regular dose and, after a clean break is achieved, cut into ¾-inch (2-cm) curds; the curds are stirred until they have the firmness of a poached egg; and only then do cheesemakers wash them.

The washing process involves drawing off a third of the volume of the cheesemaking vat in whey and replacing it with an equal amount of hot water at around 130°F (55°C). To have as little impact on the curds as possible, the hot water is poured in slowly, and the pot lightly stirred as the water is added.

The temperature of the pot will rise to the high 90s (around 37°C) after the first washing. A second washing with hot water further heats the curds, to over 100°F (around 40°C). A third and final washing brings the temperature up to about 110°F (43°C). As the curds are heated, they are stirred nonstop to release their whey.

Once the curds have firmed to the desired consistency, cheesemakers allow the curds to settle, then let the whey flow from the vats. The washed curds, knit together at the bottom of the vat, are lifted into forms and pressed into rounds of Gouda.

Waxing Cheeses

Waxing is a technique used to preserve Goudas and other hard cheeses as they age (soft cheeses subject to waxing will have too much moisture trapped within and will not age well). Waxing effectively seals off rinds from the air, restricting the growth of

Washing Gouda's rinds gives them a naturally orange hue.

surface fungi and bacteria that can influence a cheese's ripening, and allows a cheese to ripen purely from within. Cheeses with waxed rinds also age longer, keep their moist texture and do not lose weight as they age—an advantage to cheesemakers who sell their cheeses by the pound.

Traditionally, Goudas were washed with whey to keep fungal growth at bay, though now many are waxed to simplify the aging process and protect the cheeses as they travel to market. However, a more interestingly flavored Gouda with a naturally orange rind (almost the exact shade of the royal color of Holland!) can be encouraged through regular rind washing as the cheese ages.

When to Wax

Goudas can be waxed after 1 or 2 days of air-drying, post-salting, as soon as their surfaces are dry to the touch. Many cheesemakers, however, choose to age their Goudas in their caves for some time to develop flavor and firmness before they are waxed. If you choose to age your cheeses before waxing, keep the rinds of the cheeses free from fungal growth as they age by washing them with salty whey every other day until they are waxed.

Aging Gouda before waxing can also help the cheese develop the right shape.

Getting Gouda's Shape

Waxed cheeses keep their shape and do not bulge as they age. To get their waxed Goudas to have nicely rounded sides, most cheesemakers press their cheeses in Gouda-shaped forms and wax their cheeses as soon as they've dried. Gouda forms, however, are costly, and it's a challenge to find Gouda-shaped plastic containers to recycle into cheese forms!

A cylindrical Gouda, formed in a typical straight-sided cheese press, will make a straight-sided Gouda when waxed. If, however, the cheese is left to age without waxing, it will bulge slightly on its sides as it ripens, taking on the rounded shape Goudas are known for. Waxing the cheese later in its development will then preserve its rounded shape.

Waxes to Use

Cheese waxes available from cheesemaking suppliers are made of petroleum-derived paraffin made stronger and more flexible through the addition of microcrystalline wax (a component of petroleum jelly) and colored with various synthetic dyes. These refined waxes have no place in natural cheesemaking.

Beeswax is a natural alternative to petroleum-based cheese wax. Beeswax, like pure paraffin, is quite brittle, though, and may crack, especially if a cheese is aged at low temperatures or low humidity. If you choose beeswax, add a couple of tablespoons of lard, vegetable shortening, or oil to 1 pound of melted wax before dipping your cheeses to soften it up. Beeswax may be twice as expensive as paraffin-based waxes, but it contributes much more to the flavor of a cheese.

Applying Wax

The cheese must be clean and dry before waxing. If there is fungal growth or bacterial development on the rind, the rind should first be wiped clean with a brine-soaked rag, then dried with a towel, and left to air-dry before waxing.

The wax or wax-oil mixture is gently warmed in a dedicated wax pot on the stove. Be sure that your cheese will fit in the pot, and that you have enough melted wax in it to more than half submerge the cheese. Wax can smoke or burn if overheated, so keep a watchful eye on the pot as the wax melts.

To wax a cheese, hold the top half with two hands, submerge it into the wax for an instant, then remove it from the wax and hold it over the pot to let it drip. Place it to cool on a board until the wax firms, then grip it by its waxed half and submerge it again in the hot wax on its other side. Repeat this waxing and cooling until all parts of the cheese have been dipped three times.

If you have limited quantities of wax, you can also brush the wax onto the surfaces of a cheese with a natural-bristled brush.

Aging Waxed Cheeses

Waxed cheeses are much more easily aged than natural-rinded cheeses, and waxing a cheese can help a home cheesemaker have a more successful *affinage*: Once a cheese is sealed in wax it can almost be forgotten about!

Waxed cheeses do not need to be watched over in the same manner as unwaxed cheeses. They do not need to be aged in humidity-controlled environments, as the wax retains the moisture that the cheese needs to thrive. They do not need any regular surface treatments, as the wax stops the growth of surface fungi and bacteria. Waxed cheeses need only be kept in a cool environment and flipped every week as they age.

A New Take on Waxing

Waxing evolved as a means of preserving industrially made cheeses that needed protection on their long journeys to market. But waxing is no longer the preferred surface treatment for industrially made Goudas.

Much like cheddar (see chapter 21), many mass-produced versions of Gouda are now aged in vacuum packs free from air. Left in plastic to age, the cheeses' management is greatly simplified, and much work is saved.

Once ripened, Goudas are taken out of their plastic wrap and sealed with polymers. These acetate-based (acetate is an ingredient in household glue) plastic rinds are becoming increasingly common on our cheeses and are rather alarming. Found on many imported European cheeses, they can be recognized by the odd colors and strange elastic stretch of the material that covers the rind and are commonly used in conjunction with waxing and natamycin to protect the rinds of Goudas.

Natamycin, a "natural" fungicide produced by a strain of bacteria raised in bio-reactors, is applied to the rinds of many Gouda cheeses to restrict fungal growth. Natamycin's presence elicits warnings on the rinds of cheeses treated with it in some countries, but no warnings accompany cheeses treated with it in North America. Natamycin is yet another biotech ingredient that has no place in a naturally made cheese.

- RECIPE -
GOUDA/ BOERENKAAS

The name Gouda, unprotected by PDO regulations, can be applied to any cheese made in the Gouda style anywhere in the world with any quality of milk. Most of the Gouda you'll find in North America is produced on industrial dairies separated from the farm.

Boerenkaas, literally "farmers cheese," is a controlled designation for farmstead Gouda cheeses made by dairy farmers in the Netherlands from raw milk sourced exclusively from their own animals. Boerenkaas is more representative of a traditional Dutch Gouda and has considerably more character than pasteurized, industrially made versions of the cheese. If you make your Gouda with good raw milk, you can expect your cheese to be more akin to one of these.

Industrial Goudas, unlike Boerenkaas, are generally not made with raw milk. Gouda is considered a risky raw milk cheese because it does not develop as much acidity as other cheeses when it ages because of the lactose-diluting effects of the curd washing. Be sure if you use raw milk to make this cheese that the milk is of the highest quality and does not come from confined or unhealthy animals—which, of course, is a consideration you should make when making any cheese with raw milk!

INGREDIENTS

5 gallons (20 L) good cows' or goats' milk
1 cup (240 mL) active kefir or whey from a
    previous batch of raw milk cheese
Regular dose rennet (I use 1 tablet WalcoRen
    calf rennet)
5 gallons (20 L) hot water at around 130°F (55°C)
Salt

EQUIPMENT

2 5-gallon (20-L) stainless pots
Wooden spoon
Long-bladed knife
1-gallon (4-L) cheese form and press
Cheesecloth for pressing
Draining table
Cheese cave at 50°F (10°C) or less and 90% relative humidity

TIME FRAME

4 hours active cheesemaking; 3 months (minimum) aging

YIELD

Makes a 5-pound (2.5-kg) Gouda

TECHNIQUE

Slowly warm the milk to 90°F (32°C), baby-bottle-warm.

Add the kefir or whey, and mix it into the milk.

Incubate 1 hour, covering the pot and keeping the milk at 90°F to encourage the bacterial cultures to ripen the milk.

Add the rennet: Dissolve the regular dose of rennet in 1 cup (240 mL) water, and gently mix this into the milk.

Incubate 1 hour, covering the pot and keeping the milk at 90°F to encourage the rennet to set the curd.

Check for clean break. Proceed to cut the curd once it has set.

Cut the curd, in three series of cuts, into ¾-inch (2-cm) pieces. Wait a few minutes between each series of cuts to let the curds heal.

Keep the pot at 90°F while you gently stir the curds, every 5 minutes, for 30 to 60 minutes. The curds are ready when they have poached-egg-like firmness.

Pitch the curds: Allow them to settle to the bottom of the pot for 5 minutes.

Pour off about one-third of the volume of the pot's whey. For a 5-gallon batch of cheese, then, you'd pour out 1½ gallons of whey (that's 6 L of whey for a 20 L batch). Reserve some of this for preparing a washing brine.

Add back some hot water: As you stir the curds, slowly add back to the pot an equal quantity of

To make Gouda, start with basic rennet curds, and wash them with hot water several times until they firm.

Lift the curds from the pot; put them into a press; and flip them as they press to ensure an even shape.

hot water to replace the whey at about 130°F (55°C). The temperature of the pot should rise to around 98°F (37°C).

Stir the curds for 10 minutes, more quickly now and nonstop, to stop them from knitting.

Wash the curds a second time: Remove 1½ gallons (6 L) of whey and replace it with 1½ gallons (6 L) of hot water, stirring as you pour. The temperature of the pot should reach 103°F (40°C).

Stir the curds for 10 minutes to stop them from knitting.

Wash the curds a third time, removing 1½ gallons (6 L) of whey and replacing it with 1½ gallons (6L) of hot water, stirring as you pour. The temperature of the pot should reach 110°F (43°C), a temperature that your hand can just barely stand.

Stir the curds for another 10 minutes.

Pitch the curds: Let them settle to the bottom of the pot for 5 minutes.

Whey off: Pour off the whey from atop the curds. Reserve this whey for pressing the Gouda. Lightly press the curds with your hands at the bottom of the pot to encourage them to knit together.

Lift the curds into their forms: Line the cheese form with cheesecloth, and transfer the curds from the pot into the form by hand. Leave at least 1 inch (2.5 cm) of headspace at the top of the form for the press.

Press the cheese: Fill the press up halfway with warm whey, and place it atop the cheese to press it.

Flip the cheese and press again: Take the cheese out of its form, flip it, rewrap it with cheesecloth, and carefully place it back in its form to press on its other side. Increase the weight of the press atop the cheese, and press until the cheese has cooled.

Form the cheese: Once the cheese has cooled, remove the press from atop the cheese, but leave the cheese wrapped in cheesecloth in its form overnight to take its shape. Flip the cheese when convenient.

Salt the cheese: Surface-salt the cheese with 5 tablespoons (75 mL) of salt (that's 1 tablespoon for every gallon of milk [15 mL for every 4 litres]), or leave the Gouda in a saturated salt brine for 15 hours (3 hours in the brine for every pound [0.5 kg] of cheese).

Air-dry the cheese on a draining table for 1 day, flipping it occasionally to be sure that it dries evenly.

Wax the cheese (optional): Once the cheese is dry to the touch, the young Gouda can be waxed according to the instructions above. Alternatively, the cheese can be left unwaxed.

Age the cheese: Leave the waxed cheese in a cool place with a temperature below 50°F (10°C). Flip it regularly so that it ripens evenly. If the Gouda is to be unwaxed, give the cheese a regular rind washing with a strong salt brine (¼ cup salt per quart of whey [240 mL per litre]) every other day as it ages for the first month. Gouda can be eaten after 3 months of aging, but is best if ripened for at least 6 months.

CHAPTER TWENTY-ONE

Cheddar

Cheddar originated in the town of Cheddar, in the county of Somerset in the southwest of England. The most bastardized and industrialized cheese of all, cheddar's traveled an awfully long way from its origins.

Though this world-famous cheese has taken on a whole new identity (an 18th-century inhabitant of Cheddar would hardly even recognize the plastic-wrapped rubbery orange blocks bearing the name of their town as cheese), some esteemed cheesemakers continue to make cheddar according to centuries' old traditions.

West Country Farmhouse Cheddar is a PDO designation given to a select few Somerset cheesemakers who still produce cheddar according to traditional methods, which include the use of raw milk, calf rennet, and pint starter cultures, as well as bandaging cheeses with lard. A good number of artisan producers in North America who follow similar techniques also make excellent traditional versions of cheddar that are certainly worthy of their title.

Cheddaring

What makes cheddar cheddar is not necessarily its place of origin, but the particular way in which cheesemakers press the curds firm before they are formed into a cheese. This technique is known as cheddaring.

To cheddar is a verb. Cheddar cheese is cheddared. You can feel cheddared after a long day of cheesemaking. And it's the cheddaring process that gives cheddar cheeses their distinct form and flavor.

To cheddar, basic rennet curds are wheyed off, allowed to settle in the pot, and left to knit together. The curds are then sliced and stacked, and the stacks of curd restacked over and over until their consistency is right. The slices of firm curd are then ground into smaller curds, salted, stuffed into forms, and pressed into a cheese. Here are the details of the technique:

Cheddar is made in massive vats using many gallons of milk. The milk is cultured, then set with rennet. The curds are cut with cheese harps, then stirred with a large paddle; when the curds have firmed, the stirring stops and the curds are allowed to settle. A spigot on the end of the vat is opened, and the whey flows from atop the curds.

The cheddaring vat has a valley running down its center that allows the whey to flow to the spigot. The curds are pulled from the bottom of the valley up to the sides and left to knit together in two large piles. The piles of curd are sliced 2 inches (5 cm) in width and stacked four high in stacks that run the length of the vat on either side of the valley. The stacks are then turned over and over, until the slices of curd have pressed themselves firm.

Once the slabs of curd have a rubbery consistency, they are put into a curd mill and ground to a smaller size. The milled curds are put back into the vat and dry-salted to pull out their moisture (it is these salted curds that are sold as fresh cheese curds—or as the Quebecois who add them to their poutine call them,

Cheddaring the curd defines the cheddar-making process.

"crottes de fromage"—literally "cheese turds"!). The salted curds are stuffed into cheese forms and pressed with the whey remaining from the make.

Cheddaring takes a lot of milk and turns it into a compact amount of cheese with a lot of hands-on effort. The most physically demanding of all cheeses, cheddar takes a lot of work, but the results are well worth it.

Traditional Versus Industrial Cheddar

Excellent cheddar cheeses are made by hand according to traditional methods all across the world. Unfortunately, these handmade cheeses carry the same label as homogenized, mass-produced, fluorescent-orange versions of the original.

But you can't blame Kraft for ruining cheddar; the cheese was on its decline long before the infamous dairy corporation forced the closure of thousands of small cheddar factories across North America.

Cheddar: A Cheese's History

As dairy farmers in Britain began co-operatizing and mechanizing their operations in the 19th century, cheddar became their cheese of choice for large-scale production. Cheddar, once a local specialty revered across England, was subject to a redesign, featuring processes that were easily reproducible, and indeed they were reproduced across Britain and beyond.

The first cheese to be subject to the forces of globalization and mass production, cheddar's production was standardized, industrialized, and exported to all corners of the English-speaking world. The promoter of these standardized methods, Joseph Harding—who famously said, "Cheese is not made in the field, nor in the cowshed, nor in the cow, it is made in the dairy"—had no notion of terroir. His version of cheddar production, known as the Joseph Harding Method, makes the ubiquitous undifferentiated cheddar cheese blocks you're probably most familiar with. The father of the scientific method in cheesemaking, Mr. Harding transformed cheddar-making from an art into a science (whether or not that is a good thing is, perhaps, a theme of this book), setting the standard for industrial cheesemaking that has stood to this day.

Orange You Glad Your Cheddar Isn't Orange?

Orange cheddar is an unfortunate consequence of the industrialization of dairying that went hand in hand with the industrialization of cheesemaking. Originally an act of adulteration to cover up the questionable origins of mass-produced cheddar, adding orange coloring to cheddar has since become standard practice.

As dairies increased in size in the 19th century, cows started being kept in confinement and fed more readily available grain to improve efficiencies and increase their milk output. As a consequence of

Cheddared curds milled, salted, and ready to be pressed into cheddar . . . or eaten out of hand!

this new style of husbandry, cheeses lost their naturally creamy color.

Cows make colorful cheese when they feed on fresh green grass. Carotene, an essential vitamin as well as a pigment in grass, colors the milk and the cheeses of pastured cows. But confined cows do not get their daily dose of carotene in their hay, haylage, or grains, and their milk shows its vitamin deficiencies when made into cheese: A cheddar made with pastured cows' milk has a beautifully creamy color because of its carotene content; cheese made with milk from confined cattle is unnaturally white.

Cheesemakers began to add annatto to their cheddars to dress up the colorless cheeses that confined cattle produced, and that consumers refused to buy. Annatto, a natural plant dye from a South American tree known as achiote, restored the cheeses' creamy hue when added to the milk (though it didn't restore their vitamin deficiencies), and cheddar eaters thus had the wool pulled over their eyes.

When added in excess, though, annatto gives cheeses an orange glow. Some cheesemakers began adding extra color to standardize their cheeses' appearance and help their cheeses stand out from the crowd; if a little bit of annatto helps to sell cheese, they reasoned, a lot of annatto will sell even more. And by and large their reasoning was right! Today the orange color has become so much a part of cheddar's identity that many cheddar eaters won't buy the cheese if it isn't bright orange.

Plastic-Bound Cheddar

Industrially produced cheddars are sealed in plastic to age: 5-year, 10-year, and 15-year cheddars spend their long lives confined in vacuum-sealed plastic that prevents the cheeses from breathing. Only plastic-bound cheddars can be aged for such preposterously long times; traditional cheddars, bound in cheesecloth, lose moisture as they ripen and cannot be aged much beyond two years because they become too dry. Plastic keeps mass-produced cheddars moist for years and allows their ripening to continue unimpeded for decades. But what becomes of a cheese that spends a miserable decade entombed in plastic?

Clothbinding

Traditionally made cheddars are wrapped in cheesecloth and smeared with lard to prepare them for aging. Known as clothbound or bandaged cheddars, their cheesecloth wrapping is pulled from the rind before eating.

Clothbinding is a method of preserving the cheese rinds that predates waxing and plastic binding—both more modern methods of keeping rinds fungus-free. Achieving the same end as waxing, clothbinding impedes air's access to the surfaces of cheeses, preserving them for longer periods and limiting the extent of fungal influences on their development. Clothbinding also keeps cheeses' sides from bulging over time—which helps save space in the cheese cave!

To bandage a cheddar, double strips of coarsely woven cotton cheesecloth (grocery-store-bought cheesecloth is perfect for this task) are custom-cut into squares that fit (with an inch, excess) to the top, bottom, and sides of the cheddar. The pieces of cheesecloth are applied to the rinds of the cheese, then adhered with smeared lard (tallow, or salted or unsalted butter, will also do the trick). To make the process of clothbinding easier, the cheese and the fat should both be at room temperature.

When applying a clothbinding, start with the top piece of cloth, and smear it in place, then trim the corners of the cloth to make a rough circle, and adhere the excess cloth to the sides of the cheese. Next, flip the cheese and apply the cheesecloth to its bottom in the same manner. Finally, place the cheese, on its side, and apply its side binding as you roll it over the overlapped top and bottom clothbinding, smearing as the cheese turns, then trim and adhere the edges with lard.

The clothbound cheese can then be aged in a cool and humid space, and all the maintenance it will need is a regular flipping and an occasional wiping with a dry cloth to set back fungal growth on the cheese's rind.

A clothbound rind protects the flesh of a cheese from the effects of surface-ripening fungi.

Buttering the cheesecloth-covered rinds of a clothbound cheddar.

- RECIPE -
CLOTHBOUND CHEDDAR

The cheddaring process that defines this cheese is most suited to a larger-scale cheesemaking. Traditional cheddar makers took a whole day's milking of their entire herd and transformed it into one massive cheese! But seeing as you probably won't have access to the 80 or so gallons of milk that are needed for a proper cheddar, this recipe replicates the cheddaring process on a smaller scale.

This small-scale recipe for cheddar calls for 5 gallons [20 L] of milk to make one cheese, which is about the minimum. Smaller batches may be difficult to keep warm through the cheddaring process, and the smaller cheese that results will not age in the same manner nor as long as a larger cheddar. The process, which normally happens in a giant rectangular vat, is herein adapted to a large cheesemaking pot. And care must be taken to keep the curds warm and drained of their whey as they're cheddared—such pots aren't designed with this purpose in mind!

The "stirred-curd" method of cheddar-making, which is commonly found in cheesemaking guidebooks and more easily executed in a pot, produces a cheese more akin to Gouda than cheddar. There are enough questionable versions of cheddar out there that I choose to disregard this method in much the same way as I disregard big orange blocks of "cheddar" cheese as cheddar. What is distinct about cheddar is the cheddaring stage. And this recipe is about the closest you'll get to true cheddar on a small scale.

INGREDIENTS

5 gallons (20 L) good milk, preferably unskimmed
1 cup (240 mL) active kefir or whey
Regular dose rennet (I use 1 tablet WalcoRen calf rennet for 5 gallons milk)
½ cup (120 mL) good salt
¼ pound (100 g) lard or butter

EQUIPMENT

5-gallon (20-L) stainless-steel pot
Wooden spoon
Long-bladed knife
Turkey baster
1-gallon (4-L) cheese form and press
Fine cheesecloth for pressing
Draining rack setup
Coarse cheesecloth for clothbinding
Lard or butter for smearing
Cheese cave at 50°F (10°C) and 90% humidity

TIME FRAME

6 hours active cheesemaking; 6 months minimum aging

YIELD

Makes a 4-pound (2-kg) cheddar cheese

TECHNIQUE

Slowly warm the milk to 90°F (32°C), baby-bottle-warm.

Add the active kefir or whey, and mix it in lightly.

Incubate 1 hour, covering the pot and keeping the milk at 90°F to encourage the bacterial cultures to ripen the milk.

Add the rennet: Dissolve the regular dose of rennet in 1 cup (240 mL) of cold water, and gently mix it into the milk.

Incubate 1 hour, covering the pot and keeping the milk at 90°F to encourage the rennet to set the curd.

Check for clean break and a nice satisfying *pop!* Proceed to cut the curd once it has set.

Cut the curd, in three series of cuts, into ¾-inch (2-cm) pieces. Wait a few minutes between each series of cuts to let the curds heal.

Keep the pot at 90°F, and gently stir the curds for 30 to 60 minutes, to encourage them to firm up. Proceed to whey off once they have the firmness of a well-poached egg.

Whey off: Pour the whey off the pot, and reserve for pressing your cheddar.

Leave the curds to settle in the pot for 10 minutes. Keep the pot warm to encourage them to knit together. Meanwhile, remove excess whey from the bottom of the pot with the turkey baster.

Cut the curd, vertically, into 2-inch (5-cm) thick slices. Then cut the long curds in half down the middle of the pot.

Stack the curds atop one another in two or three piles.

Flip the stacks: Taking the slices of curd from the top and putting them on the bottom, flip the stacks so that all the curds are pressed evenly. Once you've finished flipping all the stacks, flip them again! Continue flipping for half an hour, keeping the curds warm all the while by turning on the stove as needed. As the curds are pressed, they will press out their whey, which will need to removed.

After half an hour of flipping, the slices of curds should have firmed up considerably and will have a strong, rubbery texture. Be sure that the curds remain warm throughout this cheddaring stage.

Grind the curds to around 1 inch (2.5 cm) in size: You can break them up with your hands

To begin the cheddaring process, let basic rennet curd knit together into a lump; cut the lump into thick slices; and stack the slices over and over to press them evenly.

When the curds have a strong and rubbery consistency, break them up into smaller curds; salt the curds; then place the curds in a cheese press to form a cheddar.

until they are roughly the right size; or use a knife or a cheddar curd grinder to achieve the same end.

Dry-salt the curds: Pour the salt over the curds in the pot, and mix them up so that all the curds are salted. Leave the salt to pull the moisture out for 10 minutes, mixing the curds occasionally so that they do not stick together, and removing excess whey as needed. Keep the curds warm during this process by periodically heating the pot.

Press the curds into cheddar: Line the cheese form with cheesecloth, then gather the still-warm curds and stuff them into the form. Apply a few pounds of pressure by balancing a quart (1 L) of warm whey in a press set atop the cheese.

Flip and press the cheddar: After 10 minutes of pressing take the cheese out of its form, flip it, and place it back in the form upside down with the aid of the cheesecloth. Reapply the press with double the weight.

Keep flipping and pressing the cheddar: Continue pressing the cheese, flipping it every 10 minutes or so, for about 1 hour, or until the cheeses have cooled.

Form the cheddar: Leave the cheese in its form overnight without a press. Flip the cheese at some point during this period to ensure that it develops an even shape.

Cure the cheddar: Leave the cheese to air-dry on a draining rack for 1 day. Flip the cheese occasionally to preserve its shape.

Clothbind the cheddar: Once the cheese has dried to the touch, the cheddar can be bandaged. Follow instructions on page 240 for bandaging the cheese. Alternatively, cheddars can also be waxed (see chapter 20 for more information on waxing).

Age the cheddar: Leave the dressed cheese to age in a cool and humid space for 6 months to 1 year. Flip the cheddar once a week, and occasionally wipe down the fungal growth on its surface with a dry cloth as it ages.

CHAPTER TWENTY-TWO

Whey Cheeses

If you make cheese, you'll be left with a windfall of whey. From a gallon of milk, you'll make about a pound of cheese yet somehow, at the end of it all, you'll still be left with almost a gallon of whey.

Whey is by far the more voluminous product of cheesemaking; it's a good idea to know the many ways to use it! In this second-to-last chapter we'll look at what to do with all the whey left over from the many cheeses of this book.

The Constituents of Whey

Essentially milk minus the curds, whey is what remains of the milk after the making of cheese. Whey contains, depending on the type of milk used and the type of cheese made, varying amounts of proteins, fats, and sugars, lactose and lactic acid, vitamins and minerals, and a healthy dose of beneficial microorganisms all suspended in the liquid portion of the milk.

The protein in whey is the albumin that does not respond to coagulation (refer to chapter 10, Paneer, for a more thorough milk protein lesson). The albumin, in its liquid form at lower-temperature cheesemaking, is unaffected by rennet and is left behind in the whey. It is this albumin that gives whey the ability to make ricotta.

Cream that has already separated from the milk also remains in the whey. The curd catches much of the milk's fat as it forms, but the fat that has already risen into cream is missed. This whey cream can be skimmed off and churned into whey butter, or left in the whey to make an especially creamy ricotta.

Whey also contains most of the milk's lactose, as well as the lactic acid transformed from the lactose by the starter cultures. Depending on the type of cheese made, as well as the lactose and lactic acid content, the whey can be either sweet or acidic.

A multitude of microorganisms remain in the whey. A portion of the cheesemaking cultures added by the cheesemaker as well as indigenous cultures of the raw milk are left in the whey at the end of a cheesemaking session. These microorganisms make whey an excellent source of cultures for starting another batch of cheese.

Many water-soluble vitamins and minerals are also left behind in whey after cheesemaking: These make whey an important source of nutrition and a valuable fertilizer.

Not all wheys are created equal, though. Because of the different methods of their making, some wheys cannot be used the same way as others.

Compare, for example, paneer whey and yogurt cheese whey. The whey left from yogurt cheese is highly acidic and contains diverse bacterial cultures of the yogurt from which it is made. The whey left from paneer-making, however, is not as acidic, and does not contain beneficial bacteria because it has been cooked to a high temperature. Yogurt cheese whey has many different uses as a result of its beneficial bacterial cultures and high acidity; paneer whey is much less useful.

Paneer whey is considered cooked whey, while whey from yogurt cheese is considered acid whey. Sweet whey, left over from rennet-based cheese-making, is the most versatile of all.

Sweet Whey

Sweet whey, remaining from cheeses made with rennet, is only mildly acidic; it contains beneficial bacterial cultures, the milk's albumin, and a portion of milk's water-soluble vitamins and minerals.

Sweet whey is the best for making ricotta, *Mysost*, and other whey cheeses because it contains the most available albumin protein. And because it is highly nutritious, and has a mild flavor and low acidity, it is also most suitable for cooking.

Sweet whey is a source of beneficial bacterial that can be used as a starter culture for another batch of cheese. Traditional methods of cheesemaking often involved saving a portion of the leftover whey as a starter culture for the next day's cheese. A technique used in the production of many traditional European cheeses, backslopping whey establishes active micro-biological communities: In reusing their whey day after day, traditional cheesemakers invigorated the cultures that helped define their cheeses. See appendix B for instructions on keeping a whey starter.

Acid Whey

The wheys left over from long-fermented and acidic cheeses such as yogurt cheese and chèvre are considered acid whey.

Acid whey has more minerals and vitamins present in it, as these elements are leached into the whey under acidic cheesemaking conditions. Acid whey is also a very rich source of beneficial *Lactobacillus* cultures and is great to use as a starter culture for cheese, as well as for other fermented foods, such as lactofermented pickles and salami. However, acid whey does not produce ricotta when cooked!

Cooked Whey

Whey left from heat-acid cheeses such as paneer, queso fresco, and ricotta has much more limited uses than sweet whey or acid whey.

Because this whey has been cooked, it does not harbor beneficial microorganisms and cannot be used as a starter culture. Cooked whey is also very low in protein, as most of the milk protein is removed by the heat-acid coagulation. Cooked whey can still improve the nutritional quality of soups, breads, and grains, although it often tastes of the acid that was used to make it—for instance, a vinegar-set paneer will give a vinegar-flavored whey.

Whey-Waste!

Most cheesemakers do not value their whey. Often considered a waste product, whey is sold to the highest bidder or simply given away.

But whey was not always treated this way. The separation of cheesemaking from the farm has created a paradigm in which whey is not valued. In the increasingly specialized world of dairying, whey has become isolated from the farms that formerly put it to many different uses.

Cheesemaking was once integrated into diverse farming operations, and leftover whey would find its way into many different facets of family farms. Specialized cheesemaking operations, removed from the farm, have evolved a different relationship with their whey: A cheesemaking factory that uses thousands of gallons of milk daily will produce thousands of gallons of whey daily, and being far from the farm, they have little use for it on-site.

Before environmental regulations limited the dumping of whey, cheese factories flushed their whey into sewers and waterways. Whey, however, is a nutritionally rich liquid, containing about one-quarter of the nutrients of milk (as measured by biochemical oxygen demand), and if it finds its way into rivers and lakes it can cause algal blooms that rob the water of its oxygen, killing fish and other aquatic life. Because of its high nutrition it is not permitted to be flushed into municipal sewers, as it can cause overloading problems downstream at sewage treatment plants.

It is a mixed blessing that whey is now considered a product of value by the processed food industry,

Cooked whey from paneer has limited uses compared with other wheys.

for there are many more sustainable uses for it, as described below. Many industrial cheesemakers now have their whey trucked away to distant processing plants where it is transformed into whey protein isolates and concentrates. Commonly found in bodybuilding supplements, meal replacements, and a growing number of food "products," whey protein now boosts the nutritional profile of many questionable foods.

The processes that concentrate whey are also energy-intensive: The high price of whey protein supplements is a reflection not of the price of the whey but of the high energy costs of concentrating and isolating the scant ounces of proteins out of gallons and gallons of whey. Produced using expensive and highly specialized equipment that separates whey's proteins, whey protein production is beyond the scope of small-scale cheesemakers.

The Many Uses of Whey

Though cheesemakers often give it little heed, whey is not to be overlooked. It is a valuable, nutritionally and microbially rich liquid that has many different uses and numerous benefits. Whey has many different applications in the dairy, on the farm, in the garden, and at home. Good whey management practices can even help cheesemakers establish a more ecologically sound cheesemaking.

In the Dairy

Whey can be used to make several different whey cheeses, including ricotta and *Mysost* (see the recipes starting on page 255), though cooked whey still remains even after the making of ricotta!

Whey can be saved from a batch of cheese to use as starter culture for the next batch. If refrigerated, whey left from the making of a cheese will remain active for up to a week, or for many months if frozen.

Whey can serve as a weight to press cheeses firm. Its warmth makes whey an excellent weight for pressing paneer, Gouda, cheddar, and Alpine cheeses.

Whey can make strong brines used to salt cheeses, or light brines to preserve such cheeses as feta and mozzarella. Whey can also be used to wash cheeses' rinds and keep them fungus-free.

On the Farm

Whey can be used as a replacement for synthetic fertilizers to replenish pastureland and hayfields that nourish the dairy animals from whose milk the whey evolves. Revitalizing the land that makes the cheese is whey's most fitting use.

Whey can be bottle-fed to young calves, kids, or lambs that aren't yet accustomed to eating grass or grain. It contains much of the nutrients and vitamins of milk; many dairies wean their young animals from their mothers with whey.

Laying hens will peck away at whey all day, while whey-fed pigs are considered fine swine. The famous Parma hams are fattened on whey from Parmigiano Reggiano cheese, and chickens find exactly the protein they need for egg development from whey's albumin. Almost any livestock's grains can be soaked in whey to help improve digestion and nutrient availability.

Like probiotics for barns, whey can also be used to help clean out sleeping areas and milking stalls to help establish bacteriologically diverse microbiomes that may keep unwanted microorganisms in check.

In the Garden

Whey can be applied directly as a liquid fertilizer to help crops grow. A rich source of nitrogen,

Whey provides excellent nourishment for many farm animals, especially pigs.

phosphorus, and potassium as well as many other micronutrients, whey is a well-balanced soil amendment that can be poured directly over the soil without dilution. Corn, squash, greens, and other heavy feeders grow beautifully with whey.

Whey can also be added to the compost pile as a nitrogen feedstock, as well as a source of the beneficial "effective microorganisms" needed to get a pile cooking. As such, whey may be one of the best ingredients to add a compost pile, because it helps achieve the high-temperature conditions necessary for good composting.

At Home

Acid whey can replace buttermilk when you make biscuits, pancakes, or sweet loaves. As an acidic ingredient it will react with baking soda to help raise all sorts of sweet and savory baked goods.

Sweet and acid wheys can replace chicken or beef broths in the making of many soups; my favorite soup, borscht, is wonderful when made with a whey base. Whey can also contribute a heartiness to stews and chilis made with it. An excellent slow-cooked meatless stew can be made with potatoes, beans, whey, and salt.

Whey can soak grains or legumes before they're cooked, to pre-ferment the grains and make them more nutritious. Grains can also be cooked in part whey, part water to help boost their nutritional profile. Whey can replace water when baking breads to give them a protein and vitamin boost.

Whey can serve as a bacteriological inoculant for sour pickles and sauerkraut. The *Lactobacillus* cultures in acid whey are the same cultures that help make lactofermented vegetables. Whey's microbiological diversity can even help get a sourdough starter started.

Whey is a fantastic organic fertilizer that grows gorgeous greens.

Sweet whey can also simply be drunk. The French call whey *petit lait* (little milk) because it tastes like a light version of milk! The Swiss turn whey left over from their cheesemaking operations into their national drink, Rivella, a sweetened and carbonated whey soda flavored with alpine herbs.

Whey can also be fed to pets: Dogs will lap it up, but cats will turn up their noses at it!

- RECIPE -
FAST RICOTTA

Italian for "cooked again," *ricotta* refers to the second making of cheese from one batch of milk: The milk is first "cooked" to make Parmigiano Reggiano or pecorino or some other Italian cheese, and the leftover whey is then "cooked again" to make ricotta.

Many of us have probably never tried real ricotta. Because the yield of ricotta from whey is very low, commercial versions of this cheese are mostly made with milk. Rare even is the artisan cheesemaker who makes real ricotta from their leftover whey: Ricotta is very energy and time consuming and requires expensive and specialized equipment to make on a large scale. And because ricotta is a perishable fresh cheese, many artisan cheesemakers are troubled with its management and sale.

Not all wheys can be used to make ricotta. The sweetest whey left over from Alpine cheeses gives the highest yield and is the easiest to strain. Ricotta made from basic rennet curd whey is creamier but is harder to strain and yields less. As for acid whey—don't even bother! Ricotta from chèvre and yogurt cheese whey forms ephemeral curds that are impossible to strain.

This version of ricotta, which I call fast ricotta, is made by adding acidity to the sweet whey in the form of lemon juice or vinegar, just as in the making of paneer. A slower and more traditional (and more delicious and better-textured) ricotta can be made by allowing the sweet whey to ferment for a day and

develop the acidity it needs to separate its remaining cheese; see the slow ricotta recipe below for details.

Ricotta is best made with several gallons of whey. You can save up whey from several cheesemaking sessions to make it if need be. However, as whey sits and refrigerates, it ferments, and ricotta made with it will be more akin to slow ricotta!

INGREDIENTS
1 gallon (4 L) fresh sweet whey, preferably more
¼ cup (60 mL) vinegar per gallon whey or ½ cup (120 mL) lemon juice

EQUIPMENT
2-gallon (8-L) stainless pot
Ladle or slotted spoon
Good cheesecloth
Stainless strainer
Large stainless bowl

TIME FRAME
1 hour

YIELD
Between 4 and 8 ounces (100–225 g) ricotta per gallon whey

TECHNIQUE
Bring the whey to a boil: Pour the whey into the pot and bring it to a boil over medium-high heat. Don't worry about stirring the whey—it will not burn on the bottom of the pot. But pay attention to the whey as it gets hot: Once it comes to a boil, it can boil over and cause quite a mess!

Pour in the vinegar or lemon juice as the whey is boiling. Don't bother stirring it; the boiling whey will mix the acid in thoroughly.

Let the whey come to a full rolling boil again, but only for a moment. The high temperatures help to ensure a full ricotta yield.

Let the whey cool for 5 minutes. Turn off the heat, and let the whey settle. You should begin to see signs of separation: The whey will become clearer and more greenish yellow, and there will be fluffy clouds of ricotta curd that have coagulated out of it. Let the whey cool for 5 minutes to help the ricotta firm up.

Strain the ricotta: Using a ladle or a slotted spoon (depending on the type of whey and the quality of milk, the curds may be easier or harder to strain), scoop up the ricotta that has risen to the top of the pot, and transfer it to a cheesecloth-lined colander perched on the stainless-steel bowl.

Let the ricotta drain and cool. Ricotta is best savored while still slightly warm. Once cooled, it should be kept refrigerated.

To make fast ricotta, bring fresh whey to a boil and add vinegar; strain the curd from the pot; and drain the curd into cheese.

– RECIPE –
SLOW RICOTTA

This even more delicious ricotta is made by allowing sweet whey to ferment at room temperature for up to 24 hours to develop the acidity that is needed to separate the curd when the whey is boiled. Neither lemon juice nor vinegar is added; this is a true whey ricotta, and the most traditional way to make this cheese.

Fast ricotta may be sweeter than slow ricotta because it is unfermented and retains its lactose, but slow ricotta develops much more interesting flavors as a result of its fermentation. The texture of slow ricotta is softer and creamier, too. However, its soft curds are more challenging to strain from the whey than fast ricotta's and take considerably longer to drain. With its daylong fermentation and lengthy straining time, slow ricotta takes considerably longer to make; but aren't all good things in life worth waiting for?

INGREDIENTS

1 gallon (4 L) fresh sweet whey, preferably more

EQUIPMENT

2-gallon (8-L) or larger stainless pot
Ladle
Good cheesecloth
Stainless strainer
Another large pot

TIME FRAME

12–24 hours

YIELD

Between 4 and 8 ounces (100–225 g) ricotta per
gallon whey

To make slow ricotta, bring fermented whey to a boil; ladle out the soft curd from the pot; and drain the curd in cheesecloth.

TECHNIQUE

Ferment the whey: Leave the whey in a closed container at room temperature for 12 to 24 hours to allow it to ferment.

Pour the whey into the pot and bring it to a boil over medium-high heat. Don't worry about stirring the whey—it will not burn on the bottom of the pot. Take the pot off the heat as soon as the whey comes to a boil so that it does not boil over.

Let the whey cool and settle in the pot for 5 minutes. You should begin to see signs of separation: The whey will become clearer and more greenish yellow, and there will be fluffy clouds of ricotta curd that have coagulated out of the whey.

Let the whey cool for 5 more minutes to help the ricotta firm up.

Strain the ricotta: The ricotta that forms will be very light and difficult to strain from the whey. To do so, ladle the contents of the entire pot through a cheesecloth-lined colander resting over a large pot.

Let the ricotta drain its whey until it has cooled. Add some salt to taste. Once cooled, ricotta should be kept refrigerated; however, if it's well salted, it can be pressed and preserved as ricotta salata.

- RECIPE -
MYSOST

Mysost is a Scandinavian whey cheese whose name means . . . "whey cheese." Commonly, and mistakenly, called *Gjetost* (meaning "goats' cheese" in Norwegian), *Mysost* is unlike any other style of cheese out there. More confection than cheese, *Mysost* has a texture and taste akin to fudge: Spread on toast, it's every Scandinavian child's dream breakfast!

Mysost is made by slowly cooking sweet whey in a process similar to the sugaring off of maple sap into maple syrup. As the whey evaporates, its fat and proteins concentrate into a paste, while the lactose sugars caramelize because of the long cooking process. The result is a cheese that is thick and chewy, caramel brown, and surprisingly sweet.

Mysost requires a vast reservoir of patience (and whey) to make well. And the whey that is used must be fresh. If it's older, it may have soured, and when cooked it can separate into ricotta. The whey is simmered slowly until it has reduced to a quarter of its volume and its color has turned caramel. Cream is then added to the mix, and the cheese is cooked more quickly and stirred nonstop for several hours until it thickens and turns a deep caramel-brown color.

INGREDIENTS

2 gallons (6 L) fresh sweet whey
1 pint (480 mL) cream
Bit of butter to grease a form

EQUIPMENT

2-gallon (6-L) stainless pot or slow cooker
Stainless bowl
Small cheese form

TIME FRAME

1 day to slow-cook the whey; 2–3 hours of constant stirring on the second day

YIELD

Makes about 1 pound (450 g) *Mysost*

TECHNIQUE

Bring the whey to a simmer over low heat or in a slow cooker.

Keep simmering the whey overnight. Be sure to leave the pot uncovered so that the liquid of the whey can evaporate. There is no need to stir the whey, as it will not burn on low heat.

Add the cream. When the whey has reduced to one-fourth of its original volume and has become a light caramel color, add the cream to the pot.

Mysost needs constant stirring for several hours to stop it from burning.

Stir the thickening cheese: Once the cream has been added turn up the heat on the pot. You will now need to stir the pot nonstop, occasionally scraping the sides of the pan, until the cheese has reached its appropriate consistency. It's finished cooking when it is a dark caramel-brown color and has a thick and goopy texture, which usually takes 2 to 3 hours.

Cool the cheese while stirring: Once the cheese has finished cooking it can be transferred to a bowl and stirred constantly as it cools to prevent it from crystallizing, much like fudge.

Form the cheese. Pat the warm *Mysost* into a buttered form, then place it in the refrigerator to cool.

Mysost will last for many months.

CHAPTER TWENTY-THREE

Cultured Butter

A cheesemaking guidebook without a recipe for butter would be like a dry piece of toast!

There are plenty of good reasons to include buttermaking in a cheesemaking book. Butter, particularly cultured butter, is very closely related to cheese; the concepts involved in making butter provide insight into the nature of milk; and if you're making cheese, you may well find yourself with leftover cream that begs to be churned into butter.

There's no better time than now to start making butter. For years, butter suffered from an image problem, forced onto it by an unfounded fear of its saturated fats. The changing winds of dietary wisdom cast doubts on the merits of butter for decades. A natural and traditional food with enormous health benefits, butter has finally shed itself of its unhealthful reputation.

But what of raw, cultured, grassfed butter? Left out of the spotlight as a result of restrictions on raw dairy, cultured raw butter from pastured animals likely packs an even more significant nutritional punch and less of an environmental footprint than the standard grain-fed confined cow, pasteurized sweetcream butter most North Americans eat. For those interested in eating butter that has real flavor, offers a probiotic and nutritional boost, and is easy to make, too, this chapter will look at considerations to help you make a better butter.

Sweetcream Butter

Sweetcream butter is churned from fresh, unfermented cream. It is the lightest tasting of all butters, and likely the butter you are most familiar with: Nearly all butter sold in North American supermarkets is sweetcream butter.

But though it's easier to find at the store than cultured butter, it's more difficult to make at home. Sweetcream buttermaking depends on a triumvirate of technologies—pasteurization, refrigeration, and centrifugation—to keep cream from culturing. Butter made without these technologies becomes cultured butter naturally: Pasteurization and refrigeration limit the growth of raw milk microorganisms that cause cream to culture, and centrifugation allows cream to be separated from milk without standing and fermenting.

Sweetcream butter also has less flavor than cultured butter. Its making is more energy intensive, as it requires pasteurizers, refrigerators, and centrifuges. And yet, despite all the added energy needed to make it, the transformation of sweet cream to butter is less efficient than changing cultured cream to butter: It takes longer to churn sweetcream butter, and all the extra time spent churning yields less! So why is this the butter we most often buy?

Industrial dairies save on capital and operating costs by churning fresh cream and forgoing the fermentation stage required for making cultured butter. Made quickly to reduce costs in large-scale dairies (time is money, after all), sweetcream butter isn't a time or money saver for the small-scale buttermaker.

Cultured Butter

Cultured butter is made with cream that is first fermented until thick. The naturally thickened cream churns into cultured butter.

Culturing the cream before churning offers many advantages to the buttermaker. Cultured cream is much easier to churn to butter than sweet cream. Before blenders and continuous churning machines made buttermaking with sweet cream a breeze, butter was first cultured, significantly reducing its churning time.

In much the same way that yogurt lasts much longer than fresh milk, cultured butter lasts much longer than sweetcream butter. The cream's fermentation reduces the lactose content in cultured butter, endows the butter with a protective population of bacteria, and gives it an acidic nature, all of which restrict the growth of unwanted microorganisms. Pasteurized unsalted sweetcream butter, like pasteurized milk, will only last for two weeks refrigerated before it begins to spoil; cultured butter can be kept refrigerated much, much longer.

Cultured butter is also an excellent probiotic: Cream cultured with beneficial cultures passes on its benefits to butter. Imagine spreading probiotics on your bread as you butter your toast! Cultured butter may also have more available nutrients and vitamins than sweetcream. Many foods are known to be more nutritious if fermented, and dairy is certainly no exception.

The culturing of the cream also yields a butter that is more flavorful than sweetcream butter. The fermentation intensifies butter's natural flavor . . . which is that much better if the cultured butter is raw.

Raw Cultured Butter

Raw cream from pastured animals is made for cultured buttermaking. Raw cultured butter tastes better, is made more easily, keeps longer, and tastes better than cultured butter made with pasteurized cream and added DVI cultures—the suggested practice for making cultured butter at home.

Before industrialization, buttermaking was a cultured process. Traditional recipes call for an intentional culturing of raw cream with cultured buttermilk left over from a previous batch of butter, while others call for letting raw cream stand and sour naturally, thanks to raw milk's indigenous microorganisms.

The diverse cultures in raw milk, combined with better farming practices involved in raw milk dairying, make raw cultured butter much more flavorful, and likely more nutritious. The celebrated cultured butters from France are still made with good raw milk (it's a mixed blessing that raw milk regulations restrict their import into North America—they'd have to travel a long way to blow us away), but only a few brands of raw butter are available commercially in North America—and only in jurisdictions that permit raw milk sales.

Raw cultured butter made from raw milk will be as safe as, if not safer to eat than, the same raw milk is to drink because of the benefits of the culturing process, which limits the growth of pathogenic microorganisms.

Whey Butter

Whey butter is a type of cultured butter made from cream left behind in the whey after cheesemaking. Whey cream, cultured by the cheesemaking process, is skimmed from the whey, fermented until it thickens, then churned into whey butter.

Cream that has separated from milk does not transform into curds because it lacks the casein protein that has reacted with rennet, and thus will remain in the whey. One of the foremost reasons to make cheese with the freshest milk is that if the milk has sat for several days, its cream will have risen and will be left behind in the whey, and the cheese that results will essentially be a skimmed milk cheese. A good way to redeem such lost cream is to turn it into whey butter.

The Cultured Buttermaking Process

Cultured butter is made from milk in a four-stage process. The cream is first skimmed from the milk; the cream is cultured until thick; the thickened cream is churned; and the butter is finished by washing, salting, and shaping. To help understand the process, here's more information on each of the stages.

Separating the Cream

Traditionally, cream was skimmed by letting milk stand and ladling the cream that rises to the top. At home, this can most easily be realized by letting raw milk sit in a wide container for several hours, either at room temperature or in a refrigerator if you prefer. The cream will form a thick layer atop the milk that can be slowly and carefully ladled off by hand.

Commercial dairies separate their cream for making butter with the aid of cream separators, centrifuges that separate the lighter cream from the heavier skimmed milk. These are relatively expensive pieces of machinery, but there are some small centrifuges available for small-scale buttermaking— however, be careful if you use them: Centrifuges can remove too much of milk's cream and leave behind a watery milk that's useless for cheesemaking.

Skimming Jersey cows' milk of its cream.

If you purchase cream for buttermaking, be wary of store-bought creams with added thickeners or stabilizers that can interfere with the buttermaking process.

Culturing the Cream

Cream can be cultured before buttermaking by adding bacterial cultures and allowing them to ferment the cream.

Many different cultures can be used to culture cream before buttermaking. In exactly the same process as making crème fraîche (see chapter 8, Kefir), kefir or kefir grains can be added to fresh cream and left to ferment until the cream is thickened. If you have good raw cream, you can even leave it to ferment on its own at room temperature without added cultures until it thickens naturally from the activity of its own native flora.

Cultured butter can also be made by backslopping with cultured buttermilk left over from a previous batch of butter—but only if raw cream is used. DVI cultures can be used to culture cream; however, they cannot be reused if you're using pasteurized milk because of laboratory-raised cultures' inherent sensitivities.

Whether it is raw or pasteurized, be sure that the cream is fresh. If raw cream is left refrigerated for too long, unwanted, cold-loving microorganisms can take hold in the milk; these unwanted cultures will flourish when the cream is left to culture and can give the cultured butter strong "cheesy" flavors. If pasteurized cream is left refrigerated too long before buttermaking, wild microorganisms will establish themselves in the cream, and when it is left to culture will give the butter an awful bitterness.

Once the cream has thickened (it is now crème fraîche), it can be churned into butter. If it is allowed to ferment longer, the butter will be even more flavorful.

Churning the Cream

Churning cultured cream causes it to separate into butter and buttermilk.

As the cultured cream is churned, very little appears to happen at first. But after a few minutes of churning, the cream suddenly separates into tiny bits of butter; with continued churning those bits of butter adhere to one another and form larger clumps that float in what's known as buttermilk.

Cream can be churned by shaking in a jar, churning in a butter churn, or blending in a blender. If you're churning in a jar, fill it only half full so that there is air inside to help with the churning. If you're using power churns or blenders, be sure to cool the cultured cream before churning—the heat developed by these tools can make the cultured butter greasy and hard to work with.

The ideal temperature for hand churning seems to be room temperature, about 68°F (20°C). If the cream is too cold, the milk fat will be too cold to coalesce and will not form butter; if the cream is too warm, the butter that forms will be soft and greasy and hard to handle.

Finishing the Butter

Once the butter has separated, it can be strained from its buttermilk by passing it through a fine sieve.

The strained butter is put into a cold-water bath and worked with cold hands or wooden paddles to massage out the remaining buttermilk and press out any air. The butter is kneaded and folded, over and over, in the cold water just like bread dough.

As the butter is kneaded, it is cleansed of its remaining buttermilk, which could otherwise shorten its shelf life. The water in which it is washed is replaced several times to be sure that there is no buttermilk remaining. Once the washing water clears, the butter can be salted.

Though cultured butter lasts longer than sweet-cream butter, it can be preserved even longer if it is salted; salting butter pulls out additional moisture, and thus increases its shelf life, just like cheese. Butter is salted by mixing in salt, leaving the butter undisturbed to allow the salt to pull the moisture out, then working the butter to expel the excess moisture.

The butter can then be shaped, if desired, by pressing it into a butter form, or scooping it into butter dishes, and patting it firm.

More About Butter

Some more bits about butter:

Butter Is Made of Fat Globules

The fat in milk exists in the form of globules, small bubbles of fat surrounded by a protective barrier, which tend to cluster in milk. The globules, lighter than the liquid milk, become more buoyant as they cluster. And as milk sits undisturbed, its fat globule clusters rise to the surface, forming a layer of cream above the milk.

The globules themselves are water repellent, but when surrounded by their protective covering, they remain in the cream. As the cream is churned, the continuous agitation and mixing of air weakens the protective barrier around its milk fat globules and encourages them to come together. As the globules come into contact with one another, they adhere, forming larger and larger globules that separate from the cream and become visible granules of butter.

Culturing the cream before buttermaking breaks down the protective covering of milk fat globules, which helps them coalesce more quickly than fresh cream's milk fat; cultured cream separates into butter in less than half the time of sweet cream. The breakdown of the protective covering of the globules also allows the globules to become more water repellent and gives cultured butter a higher fat content than sweetcream butter.

Butter Is Species-ist

Different species' milks do not separate into cream as readily. Cows' milk and buffalo milk have fat globules that cluster easily, causing these milks to quickly separate their cream, but goats' and sheep's

Balls of butter separate naturally from fermented goats' and sheep's milk when churned.

milk have fat globules that cluster less and barely separate into cream at all. Cows' and buffalo milk, therefore, are easily skimmed by ladling their cream by hand; however, goats' and sheep's milk can barely be skimmed by hand at all.

Butter is most easily made at home with cream that is easily skimmed by hand. Hence, homemade butter is not as easily made with goats' and sheep's milk, which don't readily give up their cream. Commercial goats' and sheep's butters are made from cream that is separated by centrifuge, but you don't necessarily need a cream separator to make butter from their milk.

A more traditional sheep's or goats' butter can be made without having to resort to using a centrifuge. Instead of culturing their cream with kefir, goats' or sheep's *milk* can be cultured, then fermented; and the thickened *milk* can be churned into butter. Though

goats' and sheep's cream do not easily separate from their milk, culturing the milk weakens the protective barrier around its fat globules, and when churned, the cultured milks naturally separate their butter.

The Color of Butter

The color of an animal's butter derives from the color of its cream, which, just like its cheese, is colored by the grass the animal eats and the way that animal digests the carotene in the grass. Cows and sheep leave the carotene in grass intact and pass the fat-soluble carotene along in their colorful cream, which churns a yellowy butter. Buffalo and goats convert the carotene in grass to vitamin A; their milk, their cream, and their butter are pure white, as they lack the color of carotene.

The greener the grass, the more intensely the carotene is expressed in the butter. When cows are

Goats' and cows' butter side by side—goats' milk butter is pure white.

feasting on lush and green spring grass, their butter is almost orange. But when the grass in the pastures matures and is dried by the summer sun, it loses its carotene as it sheds its vibrant colors; butter made during this time is lighter in color because the grass has lost its carotene. In winter, when animals are fed stored hay, fermented silage, or grain, the color of their butter is paler still. The difference in color between summer butter and winter butter reflects the changing nutritional quality of the animals' diet.

Industrially produced cows' butter is often colored to restore the creamy color that butter eaters expect, but that confined cattle deprived of pasture cannot produce. Confined cows' butter must be colored yellow with added colorants, much like margarine, to get the hue consumers are accustomed to; industrially produced butter's pallid color is a window into its nutritional shortcomings.

Cultured Buttermilk

A bonus from the making of cultured butter is the buttermaking by-product known as cultured buttermilk. Healthful and nourishing, thick and flavorful, and full of beneficial probiotics, cultured buttermilk is almost as valuable an end product as the butter itself. I cherish every drop of cultured buttermilk from my buttermaking—it's especially invigorating to drink after hand-churning butter!

Few of us may ever have tasted true buttermilk. Because North America eats very little cultured butter, there is very little true cultured buttermilk. Commercial "cultured buttermilk" is pasteurized, homogenized, skimmed milk cultured with one or two strains of freeze-dried DVIs. It does not have the same flavor, nor the same nutritional and probiotic benefits, as true cultured buttermilk left over from the cultured buttermaking process— particularly if raw cream or kefir culture is used. Commercial "buttermilk" also lacks the beautiful little bits of butter that float to the top of the glass. One of the great mislabelings in the dairy department, commercial "cultured buttermilk" should not carry such a name.

Sweetcream buttermaking does not yield cultured buttermilk; it yields sweet buttermilk. Lacking the acidity, flavor, and texture of cultured buttermilk, it is not sold to the public. Instead, it becomes yet another industrial additive: Sweet buttermilk is centrifuged to separate any remaining cream, then dehydrated into milk protein powder and used in the production of countless processed foods.

– RECIPE –
CULTURED BUTTER/ CULTURED BUTTERMILK

Cultured butter is much easier to make at home than sweetcream butter because it is based on traditional practices that naturally simplify the buttermaking process and is not reliant on mechanized butter churns, refrigeration, pasteurization, and centrifugation.

Cultured butter once was spread on the bread of every North American household, but the advent of industrialized dairying made sweetcream butter more available and affordable and devastated traditional farm-based cultured buttermaking practices in North America. We North Americans are now so accustomed to eating sweetcream butter that we don't even know butter is supposed to have flavor! European butter eaters, however, never adopted sweetcream butter, and many find North American butter bland.

Cultured butter is also better for baking. The breakdown of the protective covering around its fat globules during the cream's fermentation helps cultured butter expel more moisture than sweetcream butter; it therefore has a lower moisture content and a higher fat content. The reduced moisture content gives cultured butter more plasticity and workability than sweetcream butter, and results in flakier croissants and finer pastries. North American

bakers often import cultured butter from Europe because few domestic buttermakers produce an equivalent product: "European-style" butters made in North America are often adulterated with extracts of lactic cultures to give them a more buttery flavor—how uncultured!

You can use this recipe to make whey butter by substituting fresh cream with cream skimmed from whey at the end of cheesemaking session. You can also adapt it for making sweetcream butter by omitting the culturing process, and churning the unfermented cream; be aware, though, that the

To make cultured butter, churn thickened cream until it separates; strain the butter of its buttermilk; and wash the butter in cold water.

churning of the butter will take at least twice as long, and the resulting butter will have only half the shelf life of its cultured counterpart.

If you wish to make goats' or sheep's milk butter but do not have a cream separator, replace the cream in the recipe with milk, and culture it and churn it as if it were cream.

INGREDIENTS

1 tablespoon (15 mL) kefir grains, kefir, active whey, or buttermilk saved from a previous batch of cultured butter
1 quart (1 L) good cream
Good salt (optional)

EQUIPMENT

2-quart (2-L)-sized glass jar
Fine sieve
Large bowls
Ice water
Pair of wooden spoons or butter paddles
Butter form (optional)

TIME FRAME

1 day to culture the cream; 20 minutes to make the butter

YIELD

Makes less than 1 pound (450 g) cultured butter and about 1 pint (480 mL) cultured buttermilk

TECHNIQUE

Add the kefir, whey, or buttermilk to the cream in a jar. Cover the jar with a lid to keep out flies.

Let ferment, at room temperature, for about 24 hours or longer, until it thickens. If you're using kefir grains to ferment the cream, pass the thickened cream through a fine strainer to remove the grains.

Churn the thickened cream: Either vigorously shake the jar for 5 to 10 minutes or scoop the cream into a blender and blend on high for about 60 seconds. Stop the churning once the cream has visibly thickened into butter and has separated its buttermilk.

Separate the butter by pouring the butter and buttermilk through a fine sieve. Reserve the cultured buttermilk; it's a delight to drink and perfect for pancakes and can also be used as a starter culture for making another batch of cultured butter.

Cool the butter by submerging it in a large bowl half filled with cold water. Cool your hands alongside the butter so that the butter will not stick to them.

Knead the butter: Gather the bits together by hand, pressing them together into one ball. Knead the ball of butter in the water, folding it and pressing it over and over to rinse out any buttermilk left within.

Replace the water: As the butter is kneaded, it will release more buttermilk. Change the water several times, kneading the butter after each change until there is no more milkiness in the water.

Salt the butter (optional): Remove the butter from the water, and place it in a large bowl. Pour in 1 teaspoon (5 mL) of salt, mixing it into the butter with a spoon. Leave the salt to pull moisture out of the butter for 1 hour, then press the butter with the spoon to work the moisture out.

Form the butter: Press the butter into a butter form, shape it into whatever shape you please, or just pat it into your butter dish.

Unsalted cultured butter will last for several weeks if kept refrigerated. Salted cultured butter can be kept for months at room temperature.

Sourdough

I don't bake that much bread because I feel that making good cheese is easier than making good bread; and what is bread, after all, but a vehicle for cheese! But more than that, good sourdough bread is a vehicle for the fungal culture of blue cheese. So here is a recipe for starting a sourdough starter and baking a sourdough bread, in case you wish to grow your own *Penicillium roqueforti* culture from scratch!

I know that this isn't a bread-baking book, so I've decided to throw in this recipe in the appendix. And there is a lesson in this method.

The lesson from the inclusion of a sourdough technique is this: The grains from which flour derives have a diverse profile of indigenous bacterial and fungal cultures that are particularly well suited to consuming the grain if the germ within does not get a chance to grow. These indigenous cultures likely help protect their hosts from diseases by limiting the growth of foreign bacteria.

These native bacterial and fungal cultures are all you need for a sourdough starter; much as good raw milk contains all the culture cheese needs, good flour has all the culture bread needs. And the same can be said for cabbage, which has all the culture it needs to make sauerkraut, and apples, which have all the culture they need to make cider.

Starting a Sourdough

So to start a sourdough starter, get yourself some good flour with its indigenous cultures intact. Do not use supermarket brands, as these are often overtreated and do not contain live cultures (much as is the case with supermarket milk—is there a theme here?). White flour is particularly useless as a source of sourdough culture, as it is heated to a high temperature during its processing then bleached to make it white as snow; few, if any, native microorganisms survive. Whole wheat flour is a better bet; however, much commercial whole wheat flour is as processed as bleached white flour.

Stone-ground, organic, whole wheat flour is your best bet for a sourdough starter, but rye or other grains can serve as a replacement if you wish to avoid wheat. Each grain will generate a sourdough starter that has a slightly different microbiological profile.

To begin a sourdough starter, you create conditions that encourage the flour's native cultures to come to life. Regularly feeding flour and water to the culture in the jar ensures that the awakening culture is nourished. After 5 days, the starter should be active enough to bake with.

Be sure to use nonchlorinated water when you start your mother—chlorine is an antibacterial agent that can set back the development of the sourdough culture. You can dechlorinate most municipal drinking water by letting it sit at room temperature, uncovered, for 24 hours.

The method for starting a starter is as follows:

Day 1

Make a paste from a couple of tablespoons (30 mL) of good flour and a similar amount of water. Leave this paste to sit in a covered jar, on the counter, and at room temperature, for 1 day.

Day 2

Empty the jar, but don't clean it—the culture will endure on the side of the jar. Add a couple of table-spoons (30 mL) of fresh flour and water to the jar to make a new paste, and leave this to sit for a day. The old paste can be composted.

Day 3

As on day 2, empty the jar again, add a fresh flour paste, and leave it to ferment with a small amount of the old paste.

Day 4

By now, the paste should begin to show signs of active fermentation—small bubbles should be suspended within the mix. As on days 2 and 3, empty the jar and feed the starter a fresh flour paste.

Day 5

The paste should be quite active and bursting with bubbles. If it is, the starter is now ready to use for baking. If it is not, empty the jar, and feed the starter flour paste one more time.

Keeping a Sourdough Starter

Once a sourdough starter is begun, the culture needs a bit of maintenance. A daily feeding of flour and water is all it requires to stay active. A starter can also be kept in stasis in the fridge without any care for weeks to months.

If regularly fed, a sourdough starter can last for generations, but it won't necessarily get any better over time. A freshly made starter makes bread just as good as that created from an heirloom 100-year-old starter. That's because the two likely contain similar microbiological profiles originating from the flour they are fed.

A daily feeding will keep the culture active for bread-baking anytime you need it. To feed the starter, empty the jar—reserving its contents for baking bread or making sourdough pancakes—and prepare a fresh flour paste inside; only the smallest amount of residue from the starter is needed to inoculate the fresh paste with the sourdough culture. Feeding a sourdough starter is a daily ritual and can serve as a reminder to bake your daily bread. The process of feeding can be summed up as follows:

Day 1

Use the starter for baking a bread or making sourdough pancakes, leaving just a bit of old starter in the jar. Add ¼ cup (60 mL) of flour and ¼ cup (60 mL) of water, and make a paste. Cover the jar and leave it, at room temperature, to ferment for 1 day until bubbly.

Next Day

Same as day 1. Even if you don't bake every day, it's a good idea to feed your starter every day to keep it happy.

RECIPE

SOURDOUGH BREAD

If your sourdough starter is active and bubbly, it can be used to bake bread. If it has been sitting in stasis for a few weeks in the fridge or a few days on the counter, however, it will need to be fed and left to ferment for a day to reactivate it.

To make a sourdough bread, a dough is prepared with flour, warm water, salt, and some of active sourdough starter. The bread is kneaded to develop its gluten, then allowed to rest and rise in a warm place for several hours. Once the dough has risen, it's pressed to release some gas and allowed to rise again.

When the dough is ready, it can be shaped and placed on a baking sheet or into a bread pan to rise.

The bread is baked in a very hot oven to ensure a good rise. You can achieve an excellent crust by baking the bread in a cloche—a closed vessel that keeps a high level of heat and moisture in the baking environment.

INGREDIENTS

3 cups (720 mL) good whole wheat flour, plus
  ½ cup (120 mL) for flouring surfaces and hands
1 tablespoon (15 mL) salt
Active sourdough starter made with 1 cup
  (240 mL) whole wheat flour
Approximately 2 cups (480 mL) warm water
¼ cup (60 mL) coarsely ground cornmeal

EQUIPMENT

Stainless-steel bowl
Tea towel
Flat baking pan or a bread tin
Oven
Sharp knife

TIME FRAME

8 hours to prepare; 30 minutes to bake

YIELD

1 small loaf

TECHNIQUE

Make a dough: Mix together the flour, salt, and sourdough starter, then add enough water to form a soft and slightly sticky dough.

Let the dough rest, covered, for 15 minutes to help develop its gluten and make kneading easier.

Knead the dough on a floured surface, folding it in half toward yourself, pushing it down with both hands, then giving the dough a quarter turn. Keep your feet shoulder width apart, and get into the rhythm. Repeat the movements until a fluid motion develops. Continue kneading until the gluten has developed, between 5 and 15 minutes. The dough should be strong and elastic and will stretch into an almost see-through window when ready.

Let the dough rise: Cover the bowl with a tea towel and leave the dough to rise in a warm place for 4 to 6 hours. Check up on it occasionally to see how it's doing.

Press the dough. Once the dough has begun to rise, press it down to release built-up gas and encourage the natural yeasts to continue fermenting it.

Let the dough rise again: Leave it undisturbed for 1 more hour.

Shape the bread: Place the dough on a floured work surface, and press it into a neat square 10 inches (25 cm) across. Roll the square tightly into a log, and pat the ends together to make it stout. If you wish to bake a free-form loaf, sprinkle the cornmeal on a baking sheet, then lay the bread, seam side down, atop the cornmeal. If you wish to bake the loaf in a bread tin, sprinkle the cornmeal on the bottom of the tin, and place the rolled loaf inside.

Leave the loaf to rise: Lightly rub some flour onto the surface of the bread, then cover the loaf with a tea towel, and let it rise until it has visibly grown.

Prepare for the bake: Heat the oven to 450°F (230°C). Score the top of the bread with a sharp knife to a depth of 1 inch (2.5 cm) to help the bread rise right before placing it in the oven.

Bake the bread: Place the bread on its pan or in its tin into the hot oven. Leave it to bake undisturbed for 20 minutes.

Reduce the oven temperature to 400°F (200°C), and bake 10 minutes longer. The bread will be ready when tapping it on the bottom produces a hollow sound.

Let the bread cool: Take the bread out of the oven and let it cool on a cooling rack.

And then, for future blue cheese–making . . . inoculate a slice of your sourdough bread with *Penicillium roqueforti* fungal cultures.

Whey Starters

Before there were freeze-dried starter cultures, cheesemakers saved whey from one batch of cheese to use as a starter for the next. By regularly making cheese and reusing their whey, cheesemakers maintained vigorous starter cultures that gave life and flavor to their cheeses.

Fresh raw milk is not necessarily the best source of cheesemaking microorganisms. Though it does contain many beneficial cultures, raw milk may also be a source of unwanted and possibly pathogenic microorganisms, and making cheese with raw milk without adding a starter can yield unpredictable results. However, fermenting the raw milk encourages its beneficial microorganisms to thrive and creates conditions that discourage the growth of unwanted ones. A continuous fermentation created by regularly using a whey starter promotes the development of a community of beneficial cheesemaking microorganisms.

A whey starter will stay happy and healthy so long as fresh, raw milk is used in the cheesemaking process. Use older milk, or pasteurized milk, and the community of microorganisms may be disturbed, and a batch of cheese can take on unwanted qualities; the microorganisms that result in those faults could then be transferred in the backslopped whey to the next batch of cheese.

Starting a Whey Starter

A whey starter can be started in a similar manner to a sourdough starter. Raw milk, the source of all the beneficial microorganisms a cheese needs, can be handled in a similar way to whole-grain flour to summon from it a community of strong, reusable cheesemaking cultures. (A yogurt starter can be initiated in precisely the same manner.) Do not attempt to start a whey starter with pasteurized milk–just like making clabber, this technique will work only with raw milk.

To begin a starter, simply leave 1 cup (240 mL) of raw milk to sour into clabber in a covered jar at room temperature for 24 to 48 hours, until thickened.

Once thick, feed the starter again. Take a tablespoon (15 mL) of the clabber, add it to another jar, and mix in another cup (240 mL) of raw milk. Leave the inoculated milk to sit at room temperature for 24 hours until it sets.

After two to three feedings, the microbial community of the starter will be well developed, and the soured, thickened clabber can be used as a starter for a first batch of cheesemaking. The whey that flows from the curds of that first cheese is the whey starter that can be used to begin another batch of cheese. If well maintained, the whey starter can be used indefinitely.

Using/Keeping a Whey Starter

Using a whey starter is keeping a whey starter; every time a batch of raw-milk cheese is made, the culture is fed and invigorated. For best results, the starter should be fed and invigorated daily; if not fed regularly, the microbial community in the whey will change, and the starter will lose its best qualities.

A whey starter works best if left at room temperature between daily cheesemaking sessions to develop its acidity and microbial community. If cheese is not made every day, the starter will need a daily feeding to keep it vigorous.

To feed the starter between infrequent cheesemaking sessions, simply add a cup (240 mL) of raw milk to a tablespoon (15 mL) of the active whey, and allow it to ferment as if you were starting a whey starter. Keep feeding the culture daily with fresh raw milk to ensure its continued vigor. When ready to make cheese again, simply use the clabbered milk as a starter in place of the whey.

If preserved in the refrigerator, the microbial community of the starter will be preserved, but its activity will slow. The starter can be kept in the fridge for up to several months between feedings, but it will need to be re-invigorated with a feeding or two after an extended cold storage before using it for cheesemaking.

Considerations for Use

Many cheesemakers suggest keeping a separate whey starter for each different style of cheese. I would argue, though, that keeping so many starters is unnecessary; the microbial community within a raw-milk whey starter is diverse enough to develop into any style of cheese!

I often reuse whey from a mozzarella to use as a starter for a cheddar, then use that leftover whey from that cheddar as a starter for a blue. And even if I reuse the starter from the blue cheese (with a residual collection of added *Penicillium roqueforti* fungal spores) for making a Camembert, so long as I handle the Camembert cheese in the correct manner by washing its rind, the cheese will show no signs of blue.

What matters more than keeping separate starters for separate styles is keeping the starter happy and healthy. A strong and diverse whey culture can be used and reused for almost any style of cheesemaking, so long as the whey isn't cooked to a high temperature–above 110°F (43°C). A whey starter will contain a diversity of starter cultures–some mesophilic, some thermophilic–along with many ripening cultures. This diversity allows the starter to adapt to any style of cheesemaking.

Bloomy-rinded aged lactic cheeses, however, can benefit from having their own whey starters. As it can take several days of fermentation in the pot for raw milk's *Geotrichum candidum* cultures to grow, using active whey from a batch of cheese with established Geotrichum will result in a faster development of the ripening fungus in the fermenting curd.

Framework for a Natural Cheesemaking

This is a summary of best practices for making cheese naturally and a wish list for regulations that would allow such a cheesemaking to become widely practiced.

Beyond the ideals of certified organic cheesemaking, these standards ensure a traditionally and ecologically made cheese.

1. Milk

1.1. Milk shall only be used raw if animals have access to healthy pasture and are in good health.

1.2. Milk shall be pasteurized only if animals do not have access to pasture.

1.3. When not on pasture, animals should be fed dried hay and some grain, but not fermented silage or haylage.

1.4. Milk from animals undergoing antibiotic treatment or on other drugs shall not be used to make cheese.

1.5. Milk shall otherwise only be filtered before cheesemaking.

1.6. Milk shall be no older than 3 days.

1.7. Milk shall never be frozen prior to cheesemaking.

2. Culture

2.1. Cheesemakers shall keep their own cultures; they shall not source any DVI cultures.

2.2. Starter cultures to use include backslopped whey, kefir, other traditional dairy cultures, raw milk's endemic flora, and wooden tools that serve as microbiological reservoirs.

2.3. *Penicillium roqueforti* fungal spores shall be propagated on organic sourdough bread.

2.4. *Geotrichum candidum* shall be sourced from raw milk or kefir starter.

2.5. Washed-rind cultures shall be transferred from cheese to cheese on the fabric of a cloth dipped in brine.

3. Rennet

3.1. The coagulant used shall be of animal or plant origin, including calf, kid, or lamb rennet and cardoons.

3.2. Microbial rennets of any type shall only be used under special circumstances, such as for kosher or halal cheeses; microbial rennets, if used, shall not be advertised as vegetarian.

3.3. FPC rennet of GM origin is strictly forbidden.

3.4. Rennet production from farmstead milk-fed veal calves shall be permitted and even encouraged.

4. Other Ingredients

4.1. The use of calcium chloride is strictly forbidden.

4.2. Lipase shall preferably come from unprocessed rennet.

4.3. Citric acid or ascorbic acid shall not be used as an acidulant; only natural vinegars and lemon juice shall be used.

4.4. Milk powders, milk proteins, and other thickening agents shall not be used in the making of yogurt or any cheeses.

5. Tools and Materials

5.1. Cheeses shall not be in contact with plastic for more than 24 hours of their lives.

5.2. Cheese forms may be plastic, but efforts shall be made to source wooden, woven, and pottery forms.

5.3. Cheeses shall be in contact with wood or natural materials for the vast majority of their making and ripening.

5.4. Stainless steel, copper, enamel, and wood are permitted for tools, pots, and vats.

5.5. Aluminum and other materials not listed shall not be permitted.

5.6. Cheesecloths shall be made of natural materials such as cotton, linen, hemp . . . exceptions shall be made for du-rags!

6. Aging Conditions

6.1. Cheeses shall not be vacuum-sealed for aging, or for marketing.

6.2. No disposable plastic wrapping papers shall be permitted.

6.3. Efforts shall be made to age cheeses in naturally cooled environments.

6.4. Any artificial cheese cave shall be designed to be as energy-efficient as possible.

7. Rind Treatments

7.1. Brushing, brining, washing, oiling, waxing, and clothbinding are the only permitted rind treatments.

7.2. No fungicides, pesticides, or chemicals of any sort shall be applied to the rinds of cheeses.

7.3. Cheese mites shall be allowed on cheeses if consumers are made aware.

8. Cleaning/Sanitizing

8.1. Whey, water, vinegar, natural soap, and elbow grease shall be the primary cleaners used to clean tools and surfaces.

8.2. Bleach, hydrochloric acid, and any sanitizers and sterilizers shall be forbidden in the cheesemaking room.

9. Whey Use

9.1. Whey shall be returned to the farm where the milk originated to be sprayed on pastures or hayfields.

9.2. Whey may also be diverted for compost-making and to fertilize vegetable or fruit crops.

9.3. Special exemptions shall be made for whey used to make more cheese, such as *Mysost* and ricotta.

10. Legislation

10.1. Special exemptions shall allow the sale of raw milk and raw milk cheeses produced according to these standards, directly to the public, and within a locality.

10.2. Special designation shall permit the home-scale production of raw dairy products for farmers markets and other direct sales.

A Troubleshooting Guide

For each chapter:

a common problem
— The probable cause (in order of likelihood), and
  • The most likely solution(s) to the problem
    (in order of likelihood)

Chapter 8: Kefir

Kefir does not set
— Kefir grains aren't active because they haven't
  been fed in a while
— Kefir ferments slowly at low temperatures
  • Be more patient with the kefir grains, and
    make kefir more often

Kefir tastes odd or bitter
— Milk has been refrigerated too long
  before fermenting
— Kefir grains are inactive
  • Use fresher milk
  • Make kefir more often

Yogurt does not set
— Incubation temperature is too low
— Culture is not active
  • Keep a closer eye on temperature
    as the yogurt sets
  • Be sure that the yogurt culture
    is active

Yogurt too thin
— Milk was not cooked sufficiently

— Yogurt was not incubated at an appropriate
  temperature
— Starter culture inactive
  • Cook the milk at 185°F (85°C) for at least
    1 hour, stirring nonstop
  • Incubate at 110°F (43°C) for at least 4 hours
  • Use active kefir as a starter

Yogurt tastes yeasty
— Mother yogurt culture is old and inactive and
  becomes dominated by yeasts
  • Use a fresher starter

Chapter 9: Yogurt Cheeses

Yogurt cheese falls through cheesecloth
— Cheesecloth is not fine enough
  • Use finer cheesecloth

Yogurt cheese doesn't give off whey or goes moldy
— Yogurt has thickening agents such as cornstarch
  • Use a more natural yogurt

Dream Cheese in olive oil floats or grows fungus
— Dream Cheese has too much moisture
  • Salt your Dream Cheese and air-dry for 24 hours
    before rolling into balls and submerging in oil

Chapter 10: Paneer

Paneer curds do not separate from milk
— Conditions aren't right for the coagulation—
  not enough acidity and temperature too low

— Milk is UHT (ultrapasteurized)
  • Be sure that the milk is hot enough, and that enough acid is added

Chapter 11: Chèvre

Chèvre curds float
— Milk is old, or contaminated with coliforms
— Starter culture was prepared with old milk or contaminated with coliforms
  • Use fresher milk

Chapter 12: Aged Chèvre Cheeses

Aged chèvre doesn't grow fungus
— Cheese is too salty
— Cheese is too dry
  • Pay close attention to salting, air-drying, and the humidity in the cave

White-rinded cheese becomes dominated by blue as it ages
— *Geotrichum candidum* becomes dominated by *Penicillium roqueforti* because temperatures are too cold for *Geotrichum* growth
— Cheese dries out
  • Leave cheese to cure at temperatures around 68°F (20°C) and high humidity for 1 week until rind becomes well wrinkled with *Geotrichum*

Mason Jar Marcellin bloats
— Cheese too moist when packed into jars
  • Add more salt to curd, let it drain longer

Chapter 13: Basic Rennet Curd

Curd never attains clean break
— Rennet is old
— Milk is not good
— Culture is not active
  • Be sure that the quality of your ingredients is foremost

Curds fall to pieces as they are stirred
— Milk quality is poor
— Handling of curds is too rough

• Be sure that your milk is sourced from grassfed animals and is unhomogenized

Curds float or cheeses bloat
— Milk is old, and cheese becomes overwhelmed by gas-producing microorganisms
  • Use fresher milk, or if necessary, pasteurize the milk before using

Chapter 14: Pasta Filata Cheeses

Slow mozzarella doesn't stretch
— Curds are not well made
— Temperature is not high enough for stretching
  • Milk is not good enough
  • Be sure the temperature of the water bath is accurate

Fast mozzarella doesn't stretch
— Milk is too acidic, or not acidic enough
  • Try adding more or less acid next time

Mozzarella is too stretchy and floppy, and hard to handle
— Curd is too hot
  • Reduce the temperature of the water bath, or leave curd submerged for less time.

Chapter 15: Feta

Feta rinds become slimy
— Cheese was not salted before brining
— Brine is not acidic enough
  • Be sure to salt and air-dry the surfaces of the cheese before aging
  • Leave the brine to ferment at room temperature as the cheese is formed, salted, and dried

Fetas float
— Too many gas bubbles in the cheese
— Cheese was not sufficiently pressed
  • Use fresher milk to avoid gas production

Chapter 16: White-Rinded Cheeses

White rinds don't grow
— Temperatures are too low

— Cheese is too salty
  • Keep the cheese at a higher temperature
  • Use an appropriate amount of salt

**White rinds become dominated
by washed-rind ecology**
— Cheese is washed for too long, and *B. linens*
  beats *Geotrichum*!
— Cheese is too moist
  • Wash the cheeses' rinds less
  • Reduce the moisture in the cheese cave

White rinds turn blue
— Rinds not washed often enough
— Temperatures are too low for *Geotrichum*—
  Penicillium roqueforti wins!
  • Wash the cheeses' rinds more frequently
  • Age cheese at a higher temperature

"Cat fur" fungus grows
— Cheese is not salty enough
  • Use more salt during salting

Chapter 17: Blue Cheeses
Blue veins do not develop in Gorgonzola
— Cheeses not pierced enough
— Cheeses do not have enough internal spaces
— Milk not fresh enough
  • Pierce the cheeses more
  • Do not press the blue cheese

"Cat fur" fungus grows on Blue Dream Cheese
— Cheese is too moist
— Cheese is not salty enough
  • Drain the Dream Cheese for longer
    before salting
  • Use more salt during salting

Chapter 18: Washed-Rind Cheeses
Washed-rind cheeses grow blue fungus
— *Penicillium roqueforti* finds a niche in the cheese as
  a result of low temperatures or infrequent washing
  • Increase the temperature of the aging environ-
    ment, or wash rinds more often

Washed-rind cheese doesn't turn orange
— Salinity is too low
— Washing is not frequent enough
  • Add more salt to the washing brine, and wash
    rinds more often

Chapter 19: Alpine Cheeses
Curds don't knit together into a cheese
— Curds cool off before they are pressed
— Curds have become too acidic
— Curds aren't pressed hard enough
  • Press the curds while still warm, use fresher
    milk, or press the curds with more weight

Chapter 20: Gouda
Curds fall apart
— Curds were heated too quickly
— Milk is not good enough
  • Add hot water slowly, and stir the curds
    for several minutes more before the
    next washing
  • Use better milk!

Chapter 21: Cheddar
Cheddaring curds fall to pieces
— Curds are not strong enough to withstand the
  cheddaring process
— Curds aren't allowed to knit together
  long enough
  • Use better milk
  • Leave curds to sit and knit longer
    before cutting

Curds do not knit together into slabs
— Milk is too acidic
  • Use fresher milk

Cheddar grows excessively moldy rind
— Cheese was not wiped down regularly
  • Wipe down rinds once a month

Chapter 22: Whey Cheeses
Ricotta impossible to strain
— Whey is not suitable for ricotta-making

- Use whey from Alpine cheeses or basic rennet curd cheeses—whey from lactic or yogurt cheeses do not give ricotta

Chapter 23: Cultured Butter

Cultured butter does not separate
— Cream hasn't been churned long enough

— Cream is homogenized
— Cream has thickeners or stabilizers in it
  - Be patient, and use better cream

Cultured butter tastes cheesy or rancid
— Cream was old before it was cultured
  - Use fresher cream

Comparison of Microorganisms in Commonly Used Starters, Raw Milk, and Kefir

Commonly Used Single-Strain Freeze-Dried Cultures	Role in Cheese	Presence in Raw Milk	Presence in Kefir
Lactococcus lactis	Acid production	Yes	Yes
Leuconostoc mesenteroides	Gas production, flavor	Yes	Yes
Streptococcus thermophilus	Acid production at high temps	Yes	Yes
Lactobacillus delbrueckii	Acid production	Yes	Yes
Lactobacillus helveticus	Acid production	Yes	Yes
Lactobacillus casei	Acid production, flavor	Yes	Yes
Kluyveromyces lactis	Yeast for bloomy rinds	Yes	Yes
Kluyveromyces marxianus	Yeast for bloomy rinds	Yes	Yes
Debaryomyces hansenii	Yeast for bloomy rinds	Yes	Yes
Geotrichum candidum	Fungus for bloomy rinds	Yes	Yes
Brevibacterium linens	Bacteria for washed rinds	Yes	-
Penicillium roqueforti	Blue rinds and veins	-	-
Propionibacterium shermanii	Bacteria for cheese eyes	Yes	-
Penicillium candidum	Fungus for bloomy rinds	?	-

NB: Both raw milk and kefir contain many more species of bacteria, fungi, and yeasts than those included in this chart, many of which have been recently identified as contributors to the fine flavor of raw milk cheeses. Packaged freeze-dried cultures do not contain such diversity.

Though *Brevibacterium linens* and *Penicillium roqueforti* may not be present in raw milk or kefir, they are introduced through specific handling practices of cheese. *Penicillium candidum*, unlike *Geotrichum candidum* (which plays a similar ecological role), may not be a traditional cheesemaking culture, as it is not found in raw milk.

Sources

Denise Fessler et al. Propionibacteria flora in Swiss raw milk from lowlands and Alps. *Lait* 79:2 (1999): 201–09.

Hsi-Chia Chen et al. Microbiological study of lactic acid bacteria in kefir grains by culture-dependent and culture-independent methods. *Food Microbiology* 25:3 (2008): 492–501.

Jianzhong Zhoua et al. Analysis of the microflora in Tibetan kefir grains using denaturing gradient gel electrophoresis. *Food Microbiology* 26:8 (2009): 770–75.

Tugba Kok Tas et al. Identification of microbial flora in kefir grains produced in Turkey using PCR. *International Journal of Dairy Technology* 65:1 (2012): 126–31.

Karine Lavoie et al. Characterization of the fungal microflora in raw milk and specialty cheeses of the province of Quebec. *Dairy Science Technology* 92:5 (2012): 455–68.

Analy Machado de Oliveira Leite et al. Microbiological, technological and therapeutic properties of kefir: a natural probiotic beverage. *Brazilian Journal of Microbiology* 44:2 (2013): 341–49.

Simona Panelli et al. Diversity of fungal flora in raw milk from the Italian Alps in relation to pasture altitude. *SpringerPlus* 2 (2013): 405.

Lisa Quigley et al. The complex microbiota of raw milk. *FEMS Microbiological Review* 37:5 (2013): 664–98.

E. Simova et al. Lactic acid bacteria and yeasts in kefir grains and kefir made from them. *Journal of Industrial Microbiology and Biotechnology* 28:1 (2002): 1–6.

Michiko Sugai et al. Characterization of sterol lipids in *Kluyveromyces lactis* strain M-16 accumulating a high amount of steryl glucoside. *Journal of Oleo Science* 58:2 (2009): 91–96.

ANNOTATED BIBLIOGRAPHY

My Experimentation

Many of my techniques come from my own experimentations with and subsequent observations of raw milk, kefir, and the cheeses I make. My experiential self-education is, by far, the most important reference for my work.

Helpful Books

Not all of these books are cheesemaking books. Many are works on diverse subjects that have influenced my style of cheesemaking.

Wendell Berry. *The Unsettling of America*. Counterpoint, 1977. A book you'll find on the bedside table of many an organic farmer. This book presents some of the most respected philosophical arguments for small-scale, ecological growing.

Gianaclis Caldwell. *Mastering Artisan Cheesemaking*. Chelsea Green, 2012. Provides a very technical approach to the subject that can help experienced industrial cheesemakers but would likely be overly intimidating to beginners. Offers very useful and readable insight into the scientific principles of cheesemaking.

Gianaclis Caldwell. *The Small Scale Dairy*. Chelsea Green, 2014. On the subject of how to produce good-quality raw milk.

Ricki Carroll. *Home Cheesemaking*. Storey Publishing, 2002. The cheesemaking guidebook that I and many other cheesemakers originally learned from; however, it is also a book that I had to unlearn to make cheese the way I do! Provides an introduction to the industrial approach to making cheese on a small scale.

William Coperthwaite. *A Handmade Life*. Chelsea Green, 2002. For insight into the beauty of simplicity and the wisdom of traditional ways of life. An inspirational work.

Catherine Donnelly et al. *Cheese and Microbes*. ASM Press, 2014. A collection of recent scholarly articles on the ecology of cheesemaking. Of particular importance are the articles on microbial ecosystems and the traditional use of wooden tools.

David Gumpert. *The Raw Milk Revolution*. Chelsea Green, 2009. A great account of the power of the dairy-archy in controlling raw milk production, the challenges that raw milk farmers face, and the degradation of consumers' right to choose the foods they wish.

Sandor Ellix Katz. *The Art of Fermentation*. Chelsea Green, 2012. A very thorough and thought-provoking treatise on traditional fermentation practices from all around the world. Sandor spreads his admiration for and insight into almost every fermented food without prejudice or preference.

Sandor Ellix Katz. *Wild Fermentation*. Chelsea Green, 2003. Though it doesn't offer much insight into natural cheesemaking practices, its philosophies on fermentation were fundamental to the development of my cheesemaking approach. Thank you again, Sandor!

Jean-Claude Le Jaouen. *The Fabrication of Farmstead Goat Cheese*. Cheesemaker's Journal, 1990. Somewhat vague (it's not a recipe book) and geared toward the experienced cheesemaker, it is nonetheless an excellent piece on the diverse methods of making French goats' cheeses.

Anne Mendelsohn. *Milk: The Surprising Story of Milk Through the Ages*. Knopf, 2008. Though not a cheesemaking book, this is an excellent read on the nature of milk and provides historical context and instructions for making a broad selection of dairy ferments from around the world.

Paul Stamets. *Mycelium Running*. Ten Speed Press, 2005. A remarkable look into the influence of fungi on the ecology of forests, and the many ways that diverse species contribute to the health and well-being of our soil and ourselves. Features excellent information on the cultivation of forest fungi that inspired my experimentations with cheese fungi.

Tim Wightman. *Raw Milk Production Handbook*. Just a small pamphlet, but it is nonetheless full of information on best practices in raw milk production.

Technical Papers Worth Reading

Though I've consulted many papers in my work, it's rare that the papers themselves are worth reading. The following are, however:

Analy Machado de Oliveira Leite et al. Microbiological, technological and therapeutic properties of kefir: a natural probiotic beverage. *Brazilian Journal of Microbiology* 44:2 (2013): 341–49. An excellent review of findings from recent kefir studies.

Noella Marcellino et al. Diversity of *Geotrichum candidum* strains isolated from traditional cheesemaking fabrications in France. *Applied Environmental Microbiology* 67:10 (2001): 4752–59. Sister Noella Marcellino's work on *Geotrichum candidum* diversity provides an introduction to this important cheesemaking culture, unknown, for the most part, in North America.

Websites of Note

David Fankhauser's cheesemaking website offers step-by-step instructions on the making of a vast array of cheeses, as well as a guide to producing rennet from calf stomachs: http://biology.clc.uc.edu/Fankhauser/Cheese/Cheese.html.

Dominic Anfiteatro's kefir "in-site" provides an astounding compilation of information on all things kefir. His excellent and eccentric readings will make anyone a kefir lover! His work helped me understand the nature of this unique culture when I first started caring for my kefir grains: http://users.sa.chariot.net.au /~dna/kefirpage.html.

The Vegetarian Resource Group blog contains an excellent article about fermentation-produced chymosin, a genetically modified rennet that has come to dominate the cheesemaking industry with little to no public oversight: http://www.vrg .org/blog/2012/08/21/microbial-rennets-and-fermentation-produced -chymosin-fpc-how-vegetarian-are-they/.

⇒⇒⇒ INDEX ⇐⇐⇐

Note: Page numbers in *italics* refer to photographs and figures; page numbers followed by *t* refer to tables.

ABOUT THE AUTHOR

David Asher is an organic farmer, goatherd, and farm-stead cheesemaker, who lives on the Gulf Islands of British Columbia. A guerrilla cheesemaker, Asher explores traditionally cultured, noncorporate methods of cheese-making. Though mostly self-taught, he picked up his cheese skills from various teachers, including a Brown Swiss cow, named Sundae, on Cortes Island.

Asher's Black Sheep School of Cheesemaking offers cheesemaking workshops in partnership with food-sovereignty-minded organizations and communities. His workshops teach a cheesemaking method that is natural, DIY, and well suited to any home kitchen. He has been teaching cheesemaking for over seven years.

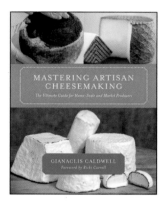
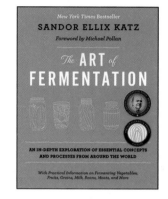
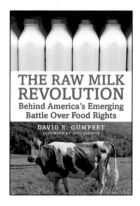

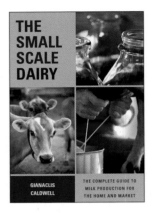